WHEN DINOSAURS ROAMED NEW JERSEY

WILLIAM B. GALLAGHER

RUTGERS UNIVERSITY PRESS
NEW BRUNSWICK, NEW JERSEY

Library of Congress Cataloging-in-Publication Data

Gallagher, William B.
When dinosaurs roamed New Jersey / by William B. Gallagher.
p.   cm.
Includes bibliographical references (p.   –   ) and index.
ISBN 0-8135-2348-6 (cloth : alk. paper). — ISBN 0-8135-2349-4
(pbk. : alk. paper)
1. Dinosaurs—New Jersey. 2. Fossils—New Jersey. 3. Geology—
New Jersey. I. Title.
QE862.D5G254   1997
567.9′1′09749—dc20   96-22044
                                                                CIP

British Cataloging-in-Publication information available

For Marjorie, Alison, Gretchen,
Kristin, and Dorothy Gallagher

CONTENTS

List of Illustrations   ix

Preface and Acknowledgments   xiii

Chapter 1. Fossils, Strata, and Time   1

Chapter 2. The Deep Past: New Jersey Before the Dinosaurs   9

Chapter 3. Who Are the Dinosaurs?   19

Chapter 4. New Jersey: Birthplace of American Dinosaur
Paleontology   27

Chapter 5. The Earliest Dinosaurs   41

Chapter 6. Heyday of the Dinosaurs   59

Chapter 7. Cretaceous Sea Life   71

Chapter 8. The Last Dinosaurs   89

Chapter 9. The Great Extinction   113

Chapter 10. After the Dinosaurs   129

Appendix A: Where to See Dinosaurs and Other Fossils in and
around New Jersey   147

Appendix B: Methods for Studying Dinosaur Footprints   149

Appendix C: How to Find Fossils in New Jersey   153

Notes   155

Glossary   161

Annotated Bibliography   167

Index   169

LIST OF ILLUSTRATIONS

1.1 Map of New Jersey's physiographic provinces 3

1.2 Geologic time chart 5

2.1 Geologic map of Precambrian and Paleozoic rocks in New Jersey 10

2.2 Paleozoic rock formations in New Jersey (chart) 11

2.3 Representative Paleozoic fossils from New Jersey 12

3.1 Saurischian and ornithischian dinosaur hip structure 23

3.2 Simplified family tree (cladogram) of dinosaurian relationships 26

4.1 Humerus and femur of *Hadrosaurus foulkii* 30

4.2 Skeleton of *Hadrosaurus foulkii* 32

4.3 Benjamin Waterhouse Hawkins's studio in Central Park, New York 33

4.4 Cope's first dinosaur 36

4.5 Operations at West Jersey Marl Company Pit near Barnsboro, N.J. 37

4.6 Charles R. Knight's painting of "Leaping Laelaps" 39

5.1 Geologic map of Newark Supergroup Triassic and Jurassic rocks 42

5.2 Newark Supergroup Triassic and Jurassic rock formations, New Jersey 43

5.3 Phytosaur skull from Fort Lee, Bergen County, N.J. 44

5.4 Cast of *Icarosaurus seifkeri*, the Triassic gliding lizard 45

5.5 *Hypsognathus fenneri* from Passaic, Passaic County, N.J. 47

5.6 Model of early bipedal dinosaur 48

5.7 *Anchisauripus* and *Grallator* footprints 49

5.8 Two *Anchisauripus* footprints headed in opposite directions 50

PREFACE AND ACKNOWLEDGMENTS

Public interest in dinosaur paleontology has surged over the past two decades as new discoveries have led to new ideas about the way in which dinosaurs lived and how they became extinct. While many of the more famous discoveries have taken place in remote and sometimes exotic locales, it is not widely known that the birthplace of American dinosaur paleontology was in New Jersey. While teaching a course about dinosaurs at Rutgers University, I found that my students were especially interested in this local connection; they were intrigued by the idea that dinosaurs could be found in their towns, and the story of the early dinosaur discoveries in the state seemed to elicit local pride. This book was written with the objective of making the story of New Jersey's dinosaurs more accessible to students of all ages. It is not only a history of the early development of vertebrate paleontology in the Garden State, it is also a summation of what we know about the paleobiology of the dinosaurs that lived in this area in the Mesozoic Era. Many of the current ideas about dinosaur origins, behavior, and extinction are presented in the text.

For a small state, New Jersey has a remarkably rich and diverse fossil record of many kinds of organisms from a wide range of geologic time intervals. So, in addition to an in-depth report on the state's dinosaurs, I have attempted to present a comprehensive picture of New Jersey's prehistoric life. I have included some practical tips for Garden State fossil collectors, and listed all the places in and near New Jersey where dinosaur specimens and other fossils may be seen. Hence this book should be of interest to anyone seeking to learn more about New Jersey's geologic past. It is, I hope, both a scholarly work on paleontology, and a practical guide to the various kinds of fossils found in the Garden State.

Many people have contributed to making this publication possible, and I can name only a few of them in this limited space. I wish to thank Peter Dodson, Donald Baird, and David Parris for their encouragement and mentoring; much of what I know about vertebrate fossils I learned from them. I am grateful to Earle Spamer, Michael Balsai, and Louis Jacobs for their helpful reviews of this work; their comments and suggestions made this a better book. I must also thank Eugene Gaffney for his review of Chapter Four and his general support of this effort. For useful discussions and camaraderie in the field, I must express my appreciation to Barbara Grandstaff, Edward Gilmore, Donald Clements, Paul Hanczaryk, Ralph Johnson, Robert Denton, and Robert O'Neill. For allowing me access to specimens in their keeping, I am grateful to Edward Daeschler, Charlotte Holton, Ralph Johnson, Bill Selden, and Mary Ann Turner. Richard Olsson and Alan Hildebrand, among others, contributed to my understanding of K/T boundary issues. Thomas Holmes provided me with useful information about some of the historical aspects of New Jersey dinosaur paleontology.

I thank my wife, Marjorie Gallagher, for her help with the manuscript preparation, for her work on the illustrations, and for her general forbearance during the course of writing this book. I must mention the patient guidance of my editor at Rutgers University Press, Karen Reeds; she initiated this work and stuck with it until completion, which at times might have seemed as long as the Mesozoic Era. I would also like to express my appreciation to several organizations and institutions who have contributed in one way or another to this project, specifically the New Jersey State Museum, the Inversand Company, the Dinosaur Society, the Rutgers Geology Museum, the Department of Geological Sciences at Rutgers, the Department of Geology at the University of Pennsylvania, and the Academy of Natural Sciences of Philadelphia. Finally, I am indebted to the many paleontologists, both professional and amateur, who have contributed to our knowledge of prehistoric life in Greater New Jersey.

WHEN DINOSAURS ROAMED NEW JERSEY

CHAPTER ONE

FOSSILS, STRATA, AND TIME

As you drive the length of New Jersey, from its southernmost point to its northwestern tip, you pass through a series of distinctive landscapes. Each one owes its special character to the geological deposits below the surface soil and to the geological forces that shaped them over the course of thousands or millions of years. From the seashore to the mountains, New Jersey's variegated scenery reflects the state's geological history. The deposits of clay, sand, and rock form a remarkably representative sample of geologic periods and contain equally good samples of the organisms that lived in those times. The progression of New Jersey's geological record is fairly straightforward; the youngest deposits are in the south and east, while the rocks are generally older and older as one goes northwest to High Point. So, if you start at Cape May Point and travel northward, you are traveling back in geologic time.

Cape May and the barrier beach islands of the coastline from Wildwood to Sandy Hook are the most recent additions to New Jersey. They are products of the last 10,000 years, built by waves and wind as sea level rose after the great glaciers of the Ice Age melted and released their water back into the ocean. This encroachment of the sea is only the most recent inundation of the ocean waters over previously exposed land; as we shall see, New Jersey has been covered by marine waters many times in the geologic past.

The Pinelands are largely underlain by the Cohansey Formation and along the northwestern oak-pine fringing forest by the Kirkwood Formation. Geologic formations are distinctive bodies of sedimentary rock that can be mapped on the basis of their composition; formations are named after geographic localities where they are well exposed. The Cohansey shales and sandstones. These are the beds of the Newark Supergroup, stretching from the Hudson River southwesterly across the state to the Delaware River, and continuing across southeastern Pennsylvania toward Maryland. These hardened sedimentary rocks were laid down in a great trough called the Newark Basin from about 220 million years ago to 185 million years ago. The redbeds are mostly devoid of fossils, but here and there they reveal traces of animals from the beginning of the age of dinosaurs. This sequence also contains gray shale, the hardened remains of the mud laid down at the bottom of large lakes; fish and aquatic reptile fossils have been found in these gray beds.

The northwestern margin of the Newark Basin is abruptly bounded by very old rocks of the Highlands province that are very much like some of the ancient rocks of New England. The hard crystalline rocks here form low mountains and hills. These are igneous and metamorphic rocks of the Precambrian age, one to two billion years old. No fossils have been found in these rocks, but younger deposits of Paleozoic sediments found in the intervening valleys do have occasional fossils in them. These fossils are the same age as the fossils found in the Paleozoic rocks in the northwestern part of the state, in the valleys and ridges of the Appalachian Mountains. The remains found in the mountain valleys are mostly of primitive sea creatures such as trilobites and brachiopods. They represent ancient ecosystems that flourished on prehistoric seafloors which have in the long course of geologic time been turned into rock and thrust up into mountain ranges.

So each part of the state has its distinctive kinds of rocks and, embedded in them, unique groups of fossils. The nature of the rocks and the fossils found in them depends upon the age of the deposits. In layered sedimentary rocks, the rocks at the bottom of the stack, deepest down in the earth, are the oldest rocks. The youngest rocks in the sequence will be at the top of the stack, closest to the surface. This concept of correlating depth with age of rock strata is known as the Principle of Superposition, and while there are some exceptions to the usual order (for instance, special cases associated with mountain building), it is the fundamental idea behind dating sedimentary rocks.

By the beginning of the nineteenth century, it was widely recognized that fossils were the remains or in some cases traces of prehistoric organisms. At about this time British geologists realized that distinctive suites of rock had characteristic fossils contained within them. Trilobites of a certain species,

Figure 1.2 Geologic time chart.

ERA	PERIOD	EPOCH	AGE (TO START OF PERIOD)
C e n o z o i c	Quaternary	Holocene	1.8 million yrs.
		Pleistocene	
	Tertiary	Pliocene	
		Miocene	
		Oligocene	
		Eocene	
		Paleocene	65 million yrs.
Mesozoic	Cretaceous		144 million yrs.
	Jurassic		208 million yrs.
	Triassic		245 million yrs.
P a l e o z o i c	Permian		286 million yrs.
	Pennsylvanian		320 million yrs.
	Mississippian		360 million yrs.
	Devonian		408 million yrs.
	Silurian		438 million yrs.
	Ordovician		505 million yrs.
	Cambrian		543 million yrs.
Pre-Cambrian	Formation of Planet Earth, approx. 4.6 billion years ago		

for example, could be found only in one or at the most several rock formations, but the vertical range of their occurrence in the sedimentary stack was definitely limited. By carefully studying and mapping the various rock formations in England and Wales, the early geologists demonstrated that unique assemblages of fossils occurred in a predictable order within stratified sequences of sedimentary rocks. This set of observations was formally called the Principle of Faunal Succession.

Around the same time in France, the famous anatomist Baron Georges Cuvier was acquiring a reputation for identifying animals from portions of their skeletons. Cuvier identified many fossil bones, and he demonstrated that fossil elephant skeletons excavated in Paris did not belong to any living species of elephant. Cuvier thus showed that animal species could become extinct, and in fact he went on to postulate that numerous groups of vertebrate (backboned) animals had succeeded each other in the rock record. He even hypothesized that a series of mass extinction events, or revolutions as he called them, had wiped out large portions of prehistoric faunas in the past. When an old assemblage of animals became extinct, a new group replaced them, only to become extinct in turn. One of Cuvier's triumphs was the identification of the "monster of the Meuse." Along the Meuse River in Belgium in 1770, ancient chalk mines yielded the well-preserved skull of a gigantic prehistoric reptile. Originally it was identified as a crocodile. When the French Revolutionary Army invaded Belgium in 1795, this skull was brought back to Paris as one of the spoils of war.[1] Cuvier studied it and pronounced it a giant sea-going lizard related to the modern monitor lizard family. He named it *Mosasaurus,* meaning "lizard of the Meuse" in Latin.

As geologists worked all over Europe to determine the vertical order of the rocks and the fossil faunas contained in them, they began to appreciate the enormous length of geologic time needed to lay down the thousands and thousands of feet of finely layered sediments now exposed as sedimentary rocks. They could use the Principle of Faunal Succession to identify equivalent rock successions separated by great distances, utilizing fossils to correlate outcrops of sedimentary rock that were separated by many miles. The Principle of Superposition allowed them to judge which rock layers were older and which were younger, relative to each other. Various estimates were made of the length of geologic time, but no direct way of determining absolute dates for rocks and fossils was available until the twentieth century. As research into the phenomenon of radioactivity progressed during the early twentieth century, scientists began to realize that the property of radioactive half-life decay could be used to provide absolute dates for minerals containing the right radioactive substances. Each kind of radioactive element emits energy at its own characteristic rate; as this radioactive energy is emitted, the radioactive element is gradually converted to a stable nonradioactive substance. This happens at a rate known as the radioactive half-life. After one half-life, one half the original amount of radioactive substance is present; after two half-lives, one quarter of the original amount is left; after three half-lives, one-eighth of the original remains; and so on. The ratio of the amount of original radioactive element to its nonradioactive product depends on time, in particular, how many half-lives have passed since the mineral crystal formed in the rock. This principle gave geologists a more accurate, absolute way of dating rocks and the fossils contained in them. Perhaps the most widely known radioactive dating techniques employs the decay of carbon-14 in organic remains, but there are various other radioisotopes that can be used for the purpose of dating rocks of different compositions and ages.

In the early 1820s, a series of fossil finds in England led to the discovery of an entirely new, previously unknown group of extinct animals. At Oxford, the Reverend William Buckland and his colleagues were studying the giant bones and teeth of a prehistoric reptile, which they named *Megalosaurus* ("giant lizard"). In the southeast of England, a country doctor named Gideon Mantell had dug up many fossils, including the remains of a large extinct plant-eating reptile, which he eventually called *Iguanodon* ("iguana tooth").

By 1842, several more such creatures had been found. A young British anatomist named Richard Owen inspected the bits and pieces and found some common features uniting these prehistoric reptiles.[2] He deduced from the structure of the hip and limb bones that these animals stood, as do modern mammals and birds, with their legs underneath their bodies, and not sprawled out to the sides, like modern reptiles such as lizards and crocodiles. Owen felt confident enough to declare that they belonged to no known branch of the reptiles: they required a new taxonomic category. He called them the Dinosauria, meaning "terrible lizards." The erect posture of the limbs was what distinguished the Dinosauria from all other reptiles.

Working under Owen's supervision, the British artist Benjamin Waterhouse Hawkins constructed enormous "lifelike" statues of *Megalosaurus* and *Iguanodon* for the Crystal Palace Exposition of 1854 in London. Hawkins followed Owen's instructions and made his dinosaurs into giant four-legged elephantine reptiles. The interior of the unfinished *Iguanodon* model was the scene of one of the most unique New Year's Eve parties in history; Owen invited twenty-one scientists to a dinner party inside *Iguanodon* on December 31, 1853. With a horn upon its nose, the quadrupedal dinosaur model looked like a gargantuan reptilian rhinoceros.[3]

There matters stood until 1858, when a famous discovery in New Jersey revolutionized our image of dinosaurs and changed the course of dinosaur paleontology.

CHAPTER TWO

THE DEEP PAST
NEW JERSEY BEFORE
THE DINOSAURS

The early geologists quickly divided up the rock exposures they were study-ing into several large categories, depending on the nature of the rocks and the fossils they contained. Some rocks were very hard and crystalline, and yielded no obvious fossils. These were generally regarded as very old, and placed in the Precambrian Era, a vast extent of time during which micro-scopic single-celled organisms were the dominant life forms on Earth. The beginning of the Cambrian Period, the first subdivision of the Paleozoic Era, is marked by the first occurrence of well-preserved and abundant fossils of multicellular animals with hard parts such as shells. This event, sometimes called the Cambrian explosion, marks the development by marine animals of protective external coverings, perhaps as a response to increasingly effective predators. At this point in geologic history, ocean waters were widespread over the lower-lying areas of the continents. Eastern North America lay close to the equator, and New Jersey was covered by a warm tropical sea.

In New Jersey, rocks of very ancient age are found in the northwestern part of the state. The New Jersey Highlands, known to geologists as the Reading Prong of the New England Province, are underlain by rocks of Precambrian age. These rocks are mostly recrystallized metamorphic rocks, that is, rocks that were once sedimentary or igneous in origin, but whose original crystal structure and composition have been changed by long burial under conditions of high temperature and pressure. Such recrystalliza-tion generally destroys any fossils that may have been present. The long-

FIGURE 2.1 Geologic map of Precambrian and Paleozoic rock outcrops in New Jersey (stippled area).

buried rocks have been exposed by uplift and erosion to produce the elevated ridges and deep valleys of the Highlands region of today.

In some of these valleys, deposits of the early part of the Paleozoic have been preserved. The Cambrian System (the name for all the rocks deposited

FIGURE 2.2 Paleozoic rock formations in New Jersey (chart).

PERIOD	GROUP	FORMATION	FOSSILS
Devonian		Marcellus	brachiopods, clams
		Buttermilk Falls	trilobites, brachiopods
		Schoharie	brachiopods, nautiloids
		Esopus	brachiopods, crinoids,
	Oriskany	Ridgeley	trilobites, brachiopods,
		Shriver	coral, nautiloids, clams
		Glenerie	crinoids, snails
	Helderburg	Port Ewen	brachiopods, trilobites
		Minisink	brachiopods, trilobites
		New Scotland	brachiopods, trilobites
		Coeymans	coral, stromatoporoids
	————	Rondout	coral, ostracodes, bryozoa
Silurian		Decker	coral, brachiopods, clams
		Bossardville	ostracodes (brine shrimp)
		Poxono Island	(no fossils)
		Bloomsburg	primitive fish
		Shawangunk	eurypterids, Arthrophycus
Ordovician		Martinsburg	graptolites, trilobites
		Jacksonburg	coral, bryozoa, crinoids
		Beekmantown	snails, conodonts
Cambrian		Allentown	trilobites, brachiopods
	Kittatinny	Limeport	stromatolites, trilobites
		Leithsville	stromatolites
	————	Hardyston	trilobites, worm burrows

during the Cambrian period of time) was first defined by Adam Sedgwick of Cambridge University as starting where the first fossils of olenellid trilobites were encountered in the lowest layer of rocks. These same fossils are found in New Jersey in the Highlands region. Trilobites are extinct members of the arthropod phylum, the same group that includes animals with jointed legs

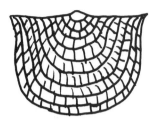

FIGURE 2.3 Representative Paleozoic fossils from New Jersey. Top left, brachiopod *Leptaena rhomboidalis,* front view; top right, crinoid stalk segment, top view and side view; bottom left, *Favosites* (honeycomb coral), top view and side view; bottom right, brachiopod *Atrypa reticularis,* side view and front view. All specimens approximately twice actual size.

like modern insects, spiders, crabs, and shrimp. Their name (trilobites) comes from the division of their bodies into three longitudinal lobes. In the Cambrian Period, trilobites were one of the most diverse and common animals in the ocean. The earliest part of the Cambrian is characterized by one group of trilobites in particular, the spiny olenellids.

In New Jersey the earliest Cambrian fossils are found in the Hardyston Formation, a quartzite unit that was originally sandstone before it was metamorphosed. Elsewhere in this formation, numerous vertical tubes can be seen in the crystalline rock; these same kinds of tubes are found in Early Cambrian rocks in many part of the world, and they are usually identified as the burrows of ancient worms living in shallow, sandy nearshore sea bottoms. While fossils of animals with hard parts are common, soft-bodied animals are rarely preserved as fossils; traces of their presence may be retained in the rock record in the form of tracks, trails, borings, or burrows, and this kind of evidence is called a trace fossil (or, in technical terminology, an ichnofossil). The trace fossils found in the Hardyston Formation are worm burrows known as *Scolithos,* and they show that what is now mountainous northwestern New Jersey was submerged under a shallow sea 540 million years ago.

Above the basal Hardyston quartzite, there is a thick sequence of limestone and dolostone, carbonate rocks that indicate deposition in a warm shallow sea. Cabbage-shaped structures called stromatolites in some of the limestone layers show the activity of mound-building blue-green algae, the first organisms to build limy reefs on the shallow ocean floor. The carbonate rocks deposited in the ancient sea's shallow banks underlie the area of the Kittatinny Valley between the Highlands to the southeast and Kittatinny Ridge to the northwest.

Kittatinny Valley is part of a much larger geological feature, the Great Appalachian Valley, which runs from Canada down to Alabama. All along the area of the Great Valley during the Cambrian Period, limy deposits built up as limestone-secreting algae contributed carbonate particles to the sea floor. Eons later, the limy soils that developed on the limestone bedrock made the Great Valley prime farming land. Primitive shelled organisms such as trilobites and brachiopods lived in the tropical waters, and their shells became buried in the limy muds. The fossil remains of these very old

Cambrian marine shellfish have been found in the limestones at Carpentersville, Peapack, Newton, and Blairstown in New Jersey.[1]

Deposition of limy sediments continued into the next geologic period, the Ordovician. The earliest Ordovician deposits in this area belong to the Beekmantown Formation, where lime was still primarily produced by carbonate-secreting algae. But, by the middle of the Ordovician, all this changed, as the primitive early algae gave way to more modern kinds of reef builders. Sea creatures that were associated with the tropical reef community of animals became abundant in the oceans of the world. Marine organisms like corals, bryozoa, crinoids, starfish, and other forms began to dominate the assemblage of marine animals inhabiting the shallow seas. More modern forms of mollusks appeared, including the complex cephalopods like the nautiloids, remote relatives of the modern squid and octopus. Snails grazed on the marine algae, and this ended the widespread fossilization of the algal "cabbage head" stromatolites. The limestone bed that entombed the first flowering of the modern reef community is called the Jacksonburg Formation. This deposit is mined for its pure lime, which is used in Portland cement. This is how ancient seabeds become modern buildings!

Toward the latter part of the Ordovician Period, oceanic conditions changed greatly in this area. All along the Appalachian Mountain front, we can see a change upward in the geologic section from the basal beds of Cambro-Ordovician carbonates in the Great Valley to the dark gray shales and slates in the foothills of the Appalachian ridges. The great change-over in sediment type indicates that the shallow sea deepened to oceanic depths, and limestone deposition ceased. Volcanic activity spread ash across the deepening seabed; this ash is preserved today as thin layers called bentonites.

Fossils of deeper-water animals are found in the gray shales and slates of the late Ordovician Martinsburg Formation. In addition to small brachiopods and trilobites, one of the more common fossils in some late Ordovician rocks are curious creatures called graptolites. Graptolites were floating colonies of marine animals that were widely distributed across Ordovician oceans; this means that graptolite fossils are good index or zone fossils that can be used to indicate a certain level of rocks across a wide area. Graptolite fossils are usually found as small saw-toothed branches of a colony that had been attached to a larger central bulbous float. Their microscopic anatomy indicates that they were distant relatives of backboned animals; the first true vertebrates appeared in Late Cambrian time, and by the Ordovician, primitive jawless fish were spreading through freshwater rivers and streams.

Deposition of marine sediments, in the form of shallow-water limestones and deeper-water shales, was terminated at the end of the Ordovician Period by a period of mountain building which geologists have named the Taconic orogeny. Evidence for this episode is contained in the coarse sediments called conglomerates that accumulated in this area in the early Silurian Period. These coarse conglomeritic rocks are a very hard natural concrete; pressurized and tilted upward in later mountain-building episodes, the eroded edge of this great sheet of rock today forms the crest of the Kittatinny Mountains, extending north into New York as the Shawangunks and southwest across Pennsylvania as Blue Mountain. This rock unit is known as the Shawangunk Formation, and it appears to have been deposited as pebbles and coarse sand in steep, swift-flowing streams running off the slopes of high, young mountains. Today, the Delaware River cuts through the hard ridge of the Shawangunk Formation at the Delaware Water Gap, exposing the dramatically tilted beds of early Silurian age.

Interbedded with the coarser rocks, especially in the middle of the formation, are thinner layers of dark shale. These seem to represent estuaries or embayments, tidal fingers of the sea that had retreated to the west. In the dark shales are fragments of fossils from animals known as eurypterids. Sometimes called "sea scorpions," these large arthropods' closest living relatives are the horseshoe crabs. Like horseshoe crabs today, eurypterids may have come into bays and estuaries to spawn. The concentrations of eurypterid fossils found in the Water Gap area are all fragmentary and small, suggesting the shed skins of growing juvenile eurypterids. Larger and more complete sea-scorpion specimens are known from Silurian deposits in New York; some fragments from the Buffalo area indicate that eurypterids may have grown up to nine feet long, making them the largest arthropods of all time.

The top of Kittatinny Ridge is the route for the Appalachian Trail from the Delaware Water Gap to High Point. Along this stretch, the trail follows the crest of the ridge formed by the very hard erosion-resistant conglomerate of the Shawangunk Formation. Some of the slabs of rock along the trail expose a very odd surface; the bedding plane of the slab is covered with a horizontal criss-crossing mat of ridged rootlike tubes. The precise nature of the organism that made these trace fossils is unknown. Some authorities think that the tubes were made by animals of some sort, while others see in them evidence of early land plants. These problematic fossils are called *Arthrophycus* ("jointed plant"), and sharp-eyed trail hikers may run across them while picking their way across the hard rocky slabs of Kittatinny Ridge.

Along the northwestern side of Kittatinny Ridge the hard conglomerates and sandstone are replaced by the red and green beds of the High Falls Formation. These shales and siltstones of Middle Silurian age were deposited by slower-moving rivers and streams flowing over a flatter coastal-plain environment. In some of the reddish stream deposits, the remains of early freshwater fish accumulated and were preserved. These do not look particularly impressive today; the largest of these fossils is less than two inches long, and they appear to be featureless oval pebbles until observed under a hand lens or low-power microscope. Then the seemingly smooth surface reveals a pattern of curving microscopic ridges, looking somewhat like fingerprints. This identifies the oval structures and smaller scattered fragments as the remains of ostracoderms ("shell-skin"), primitive armored jawless fish. These little freshwater fish had heads completely encased in a single bony head shield, with a hole on the bottom for the mouth, and the tail sticking out of the back of the bony shield. They were probably bottom feeders, swimming along the river beds and sucking up the organic detritus.

Above the High Falls redbeds are a series of calcareous rocks consisting of carbonates (limestone, dolomites) and limey shales and sandstones. These Late Silurian deposits indicate the return of a warm shallow sea environment to the Valley and Ridge region. Since the Late Silurian rocks are weak, and not very resistant to erosion, they are found underlying the valley of the Big Flatbrook in Sussex County. Some of the rocks exposed along Big Flatbrook display ripple marks and mud cracks, showing the influence of shallow tidal currents, and the intermittent drying out of the marine sediments, perhaps on the edge of a great subtropical tidal flat. Fossils are few in the Poxono Island Formation and the Bossardville Limestone, but the latest Silurian Decker Formation contains abundant remains of marine invertebrates. In particular, the exposures of Decker limestone outcropping in the basal part of Walpack Ridge in Sussex County herald the reappearance of the reef ecosystem in the Late Silurian sea. Reef animals were decimated during a mass extinction event at the end of the Ordovician Period, but by Late Silurian time corals were growing in profusion again on the warm limey shallow sea bottom. Accompanying the corals were brachiopods, crinoids, bryozoa, and stromatoporoids, spongelike mound builders that grew layered colonies on the ocean floor.

Today the prehistoric reefs form low ridges and mounds that stand up from the surrounding landscape on both sides of the Delaware River north of the Delaware Water Gap. From late Silurian time until well into the next period of geologic time, the Devonian, the upper Delaware Valley was covered by warm shallow seas. In some areas corals grew in great masses during the earliest part of the Devonian; in other places stromatoporoids were the major reef builders. No matter which reef-building organism dominated, the reef ecosystem itself was a diverse community of varied animals. Around the mounds of the reef grew gardens of crinoids, sometimes called sea lilies because of their flowerlike appearance; they are actually relatives of the starfish. In some layers of early Devonian strata their disjointed circular columnals are the chief rock-forming mineral. Among these crinoids, a lower tier of stalked brachiopods filled in the spaces between the sea lillies' rooted stalks. Brachiopods are bivalved shellfish that superficially resemble clams; however, their internal anatomy, especially their feeding apparatus, is completely different from the soft parts of clams. They still survive today, in harsher marine environments where most other immobile marine organisms will not live. They are found, for example, clinging to the rocky walls of cold, current-swept fjords in places like Alaska or New Zealand. Back in the Devonian Period they were very common in warmer waters, on soft sea bottoms, where they grew profusely around the ancient reefs. Trilobites, those now-extinct arthropods that are among the first common fossils of the Cambrian, were also present, crawling around the sea floor looking for tasty morsels to eat. Clams and snails were about, but they were not nearly as abundant as the brachiopods. In the sea water over the reef swam nautiloids, ancient cousins of the pearly nautilus. These came in two forms, straight-shelled and coiled.

Sea level rose and fell, and the marine waters covering northwestern New Jersey deepened and shoaled. Beds of shale, limestone, and siltstone preserved the profuse sea life of the Devonian waters.[2] Toward the later part of the middle of the period, things began to change again. An influx of silt and mud flooded the clear bottom waters, as sediments were dumped into the shallow sea from rising highlands to the east. Mountains were growing again, probably as a result of the approach of Europe, which collided with North America toward the close of the Devonian in an episode known as the Acadian orogeny. Mountain chains originate during the process of continental drift, as continental blocks collide, crumpling up ancient horizontal seabeds into folded and faulted mountains. In this case an ancient ocean bed was compressed and folded upward by the collision of the New England region and the Canadian maritime provinces with the slowly moving block of northwestern Europe, constructing the old mountains of New England. This mountain-building event dumped tremendous quantities of sand and gravel into the area of northeastern Pennsylvania and southeastern New York, creating the beds of the Pocono front and the Catskill Mountains.

Again the sea retreated before the rising land, and thick deposits of red sandstone and conglomerate were laid down in the prehistoric deltas formed by the rapidly flowing rivers that drained the steep new mountains and brought down more loads of sediment. In this environment, something new was stirring on land for the first time: backboned animals were making the transition from a water-dwelling to a land-dwelling way of life. In time, still distant in the future, their descendants would spread throughout the land and grow into many fantastic forms, some huge, some ferocious. They would become the dinosaurs.

CHAPTER THREE

WHO ARE THE DINOSAURS?

In the later part of the Devonian Period, forests began to flourish on land. Evidence for large trees is found in the Catskill redbeds of New York. To be sure, small primitive plants probably began to spread over the land in the Silurian, and by the early Devonian there were several kinds of terrestrial plants that grew over the land's surface by means of simple creepers and runners, like ground-covering vines today. But the forests did not reach their full size until the late Devonian. By this time there were abundant arthropods on the land; spiders and insects roamed these first forests.

In the waters, a new type of fish had appeared. The freshwaters were dominated by crossopterygians, fish with fleshy fins that were supported by stout bones, rather than the thin bony rays in the fins of most of the fish with which we are familiar today. These lobe-finned fish were the dominant animals of freshwater streams, rivers, lakes, and ponds. Their remains have been found in the late Devonian Catskill beds of Pennsylvania, where they lived in the shallow murky waters of the streams pouring off the rising Acadian Mountains.

The coastal plain of the Catskill beds was also home to another newcomer. In 1993, fossil bones found in the late Devonian redbeds of Pennsylvania have been identified as the remains of a very early amphibian, a terrestrial tetrapod. *Hynerpeton bassetti,* as this early land vertebrate has been named, had a flat head and stout strong shoulders for walking across the muddy swamps of the Catskill delta 365 million years ago. This is the second oldest known amphibian fossil, the oldest amphibian bones having been found in Scotland.[1]

During the Carboniferous Period (subdivided in America into the earlier

Mississippian and the later Pennsylvanian Periods), which followed the Devonian, the great forests spread across the tropical deltas, growing lushly in swamps where their remains would turn into thick beds of coal. Amphibians of many sizes and shapes inhabited the coal swamps, including the giant *Eryops,* an alligatorlike creature that grew to ten feet in length. As the coal swamp trees fell and buried each other, they were turned into peat. Under the compaction and heating of later mountain building, these peat beds became seams of coal such as those found in the anthracite region of northeastern Pennsylvania. New Jersey was probably a mountainous region at this time, as a result of the Acadian mountain-building episode.

It was in the tropical coal swamps of the Pennsylvanian Period that the remote ancestors of the dinosaurs first appeared. The skeletons of primitive reptiles are found closely associated with plant fossils of the coal beds. In fact, some of the oldest reptiles now known were found inside the petrified stumps of hollowed-out coal forest trees in Nova Scotia. These animals were probably trapped and buried in the hollow stumps. They have been given the name *Hylonomus lyelli.* Early reptiles were small, lightly built creatures with pointed teeth. They probably arose from primitive amphibian stock.

There was plenty for the early amphibians and early reptiles to eat in the ancient coal forests. Most amphibians and early reptiles appear to have been carnivores, judging from their sharp teeth. But if all these early land vertebrates were eating animal protein, then whom were they preying on? Today, land predators eat plant-eating vertebrates, in the relationship that has become known as the food chain (although "food web" is probably a better analogy for the more complex relationships between eaters and eaten). In the Pennsylvanian swamps, not too many early tetrapods seem to have been vegetarians. The chief source of food for these land predators was probably the large insects and other giant arthropods that were feasting on the abundant plant food and turning it into animal protein. Giant dragonflies with wingspans of two feet and foot-long cockroaches flew and scuttled through the coal swamp. They would have made more than adequate dinner fare for the small, sharp-toothed reptiles and the larger amphibians that inhabited the drier areas of the swamp.

The chief difference between amphibians and reptiles lies in how well-adapted to life on land the two groups are. While amphibians can breathe air and get about on land with their legs and feet, most amphibians must lay their soft eggs in the water and spend part of their life cycle as aquatic creatures. So they never totally lose their dependence upon the watery environment from which their fishy ancestors came. But reptiles developed the hard-shelled egg that provided a protected mini-environment for the developing reptilian embryo, buffering the embryo from the harsh conditions of dry land. Reptiles also had a stronger skeleton and better limbs for getting around over the higher and drier portions of the terrestrial realm.

Very early on, the primitive solid-skulled reptiles gave rise to two main branches of reptilian evolution. One branch was characterized by having a single pair of openings, low and in the back of the skull; this group is known as the synapsids, and among them are the ancestors of mammals. The second group had two pairs of openings in the back of the skull behind the eyes, and this group, called the diapsids, includes the lizards, snakes, crocodiles, and dinosaurs.

Initially, the synapsids flourished during the last period of the Paleozoic Era, the Permian. A wide variety of plant-eating and meat-eating mammal-like reptiles developed increasingly specialized jaw mechanisms with teeth that were differentiated along the tooth row according to their function. By the end of the Permian, the advanced cynodont synapsids had begun to evolve incisors, canines, premolars, and molars. But as the Permian Period came to a close, a series of extinctions swept through the populations of synapsids, decimating their ranks. At about the same time, there was a great dying in the marine realm and many oceanic families became extinct (such as the trilobites, the lords of the early Paleozoic sea). This major break in the fossil record marks the boundary between the Paleozoic and Mesozoic Eras, and the profound effects of the Permo-Triassic mass extinction, as scientists call this event, reorganized the nature of life on earth. Some geologists have suggested that this mass extinction was caused by the drifting together of all the earth's continents, producing mountains, draining away the seas, and affecting climate. In the ocean, mollusks became the dominant marine invertebrates. On land, synapsids were still present, but their importance was increasingly overshadowed after the extinction by a group of diapsids known collectively as archosaurs.

At the beginning of the age of reptiles in the Triassic Period, the archosaurs began to diversify into a variety of groups. All of these animals were related by the possession of additional holes in the skull, which made the bone of the head light and flexible, while offering more area for the attach-ment of larger jaw muscles at the same time. A parallel lineage to the archosaurs consisted of the lizards and their relatives, including various ex-tinct forms. Among the archosaurs, a more advanced assemblage is some-times recognized as the thecodontomorphs. While not closely related, they held in common the way in which their teeth are secured in the jaw. Unlike most lizards, thecodont reptiles have teeth set in well-developed sockets in the jaw. They frequently shed their teeth while new ones grow in to replace the lost dentures. Crocodiles and alligators have thecodont teeth; so did dinosaurs.

But this is getting ahead of our story. In the middle of the Triassic, a number of different archosaur stocks arose. Some were small armored types that probably ate small animals like insects or other arthropods. Another group became large four-footed carnivores like *Postosuchus,* which grew up to eighteen feet long; these archosaurs, known as the rauisuchids, were the biggest predators on land during the Triassic. Still another stock of arch-osaurs went back to the water and adapted an aquatic mode of life; these animals are called phytosaurs, and they resembled crocodiles in appear-ance and habits. A fourth line gave rise to small, lightly built agile creatures that paleontologists call ornithosuchians ("bird-crocodiles"). Some of these animals apparently evolved the ability to stand upright on their hind legs, and it is this group that scientists think produced the ancestors of the dinosaurs.

Lagosuchus ("rabbit-crocodile") is one candidate for a close relative of the dinosaur ancestor. This small (one foot in length), slender reptile's fossil skeleton displays some features that are very dinosaurian in nature. Besides having similarities in the ankle and wrist bones, this little ornithosuchian has a hip structure that is comparable to the hip structure of dinosaurs.

This is really what defines a dinosaur. All dinosaurs share a distinctively shaped hip, composed of six bones (three on the left side, three on the right), which enclose an open socket called a perforated acetabulum. The open acetabulum is a socket for the ball joint of the top of the thigh bone, which in dinosaurs fits neatly in the space in the middle of the hip. This arrangement allowed early dinosaurs to do something totally new among land animals at that time; it allowed them to stand upright and go about on their hind legs in an erect posture. Although some later dinosaurs would go

Saurischian Pelvis

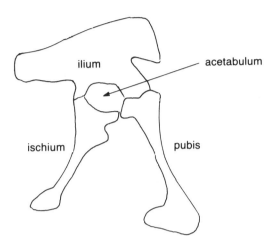

Ornithischian Pelvis

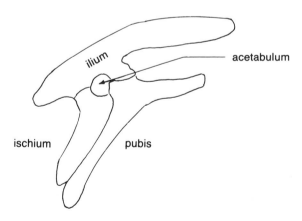

FIGURE 3.1 Saurischian and ornithischian dinosaur hip structure
(from W. B. Gallagher, *Dinosaurs: Creatures of Time,* New Jersey State Museum
Bulletin No. 14 (1990), Trenton, N.J.).

back to walking around in a four-footed stance, they still maintained the erect posture by keeping their legs in underneath their bodies rather than splayed out to the sides. All modern reptiles, such as turtles, lizards, and crocodiles, move about with their legs sprawled out to the sides, their limb joints bent at a right angle. The upright position of the limbs is what distinguishes dinosaurs from living reptiles. Dinosaurs also developed a distinctive ankle structure, an anatomical feature shared with their close cousins, the flying reptiles or pterosaurs.

By the late part of the middle Triassic, about 230 million years ago, the first dinosaurs appeared. They are mostly smaller two-legged carnivorous forms like *Eoraptor* and *Staurikosaurus,* known from late middle Triassic beds of South America.[2] By the late Triassic, lightly built bipedal predatory dinosaurs had spread throughout the supercontinent of Pangaea.

Early on in their development, however, the ancestral dinosaur stock split into two lines. One lineage continued the basic generic small carnivore model, and from this branch arose all later predatory dinosaurs both large and small, as well as the family of large heavy-bodied, long-necked, long-tailed plant-eaters called sauropods. The other group diversified into a wide variety of herbivorous dinosaurs, including the plated dinosaurs, the armored dinosaurs, the bone-headed dinosaurs, the horned dinosaurs, and the duck-billed dinosaurs. The chief difference between the two groups of dinosaurs lies in the bony structure of the hips. The saurischian dinosaurs, including all the meat-eating dinosaurs and the sauropods, shared a hip structure in which a forward-pointing bone called the pubis formed a prong at the front of the hip, a structure seen in other reptiles. Another shared feature of the saurischians was an asymmetrical grasping hand. In the other group, the ornithischian dinosaurs, the hip structure was more birdlike and the pubis bone was rotated around to parallel the rear-pointing bones of the hip (see figure 3.1). In addition, the ornithischian dinosaurs possessed an extra bone on the front of the jaw, a toothless projection called the predentary bone, which gave their lower jaws a beaked appearance.

But both kinds of dinosaurs shared the open hip socket that positioned their legs in under their bodies. With the vertical placement of the limbs under the mass of the body, the dinosaurs could walk and run in a gait that allowed the legs to swing fore and aft, instead of out to the sides in the typical reptilian sprawl. In this stance the early dinosaurs could rear back on their hind legs, standing and walking on two feet. Some of the later larger dinosaurs returned to a primarily four-footed posture, while retaining the ability to rear back and stand on the hind legs when necessary; they were what scientists would call facultatively bipedal. Still other dinosaurs became so heavy-bodied and their centers of gravity so low that they became strictly quadrupedal, and these forms lost the ability to stand on their hind legs alone. Examples of this purely four-footed type are the horned dinosaurs or ceratopsians and the armored dinosaurs known as ankylosaurs.

All dinosaurs were land animals. Some popular books on dinosaurs depict large extinct flying or swimming reptiles as dinosaurs. This simply is not accurate. While it is true that flying reptiles seem to be closely related to dinosaurs, they lack the characteristic details of bony anatomy that distinguish true dinosaurs from all other animals. The marine reptiles of the Mesozoic Era belonged to several different groups, none of which were particularly close to dinosaurs. While some dinosaurs may have taken an occasional dip in rivers or lakes, they were primarily terrestrial animals; paleontologists no longer think, for example, that the large sauropods spent most of their time buoyed up by water in streams and lakes. They were probably dinosaurian giraffes that browsed off the leaves of higher trees, wandering across the landscape in search of fresh foliage to eat.

But while some dinosaurs grew to large size, like the great sauropods, there were many dinosaurs in the small- to medium-sized category. As we have seen, the early dinosaurs were small, and many of the predatory dinosaurs throughout the Mesozoic were of modest dimensions. Dinosaurs came in all sizes as well as many different shapes. They were a very diverse group of animals, inhabiting a wide range of land habitats. Among the ornithopods, there were many small and large forms, all plant-eaters, including duck-billed dinosaurs such as *Hadrosaurus.* The thyreophorans are comprised of the plated dinosaurs like *Stegosaurus,* and the armored ankylosaurs. The marginocephalians include the bone-headed pachycephalosaurs and the horned ceratopsians. Dinosaur diversity and relationships are summarized in figure 3.2, a kind of family tree known as a cladogram.

Our knowledge of dinosaurian variety is constantly increasing. For example, there are now six families of sauropods recognized by paleontologists in the fossil record. New discoveries are changing the way we look at

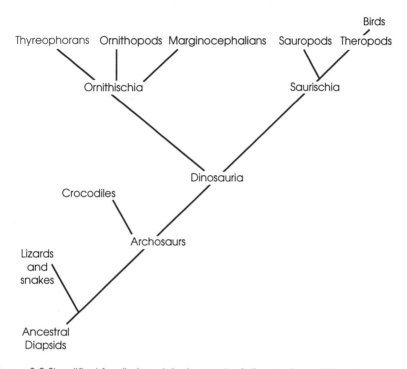

FIGURE 3.2 Simplified family tree (cladogram) of dinosaurian relationships.

dinosaurs, and every fresh find of a new specimen of dinosaur has the potential to revise the way in which one or another group is classified. Since we are in a very active phase of dinosaur discoveries right now, that means that dinosaur relationships are in a state of reevaluation and alteration.

CHAPTER FOUR

NEW JERSEY
BIRTHPLACE OF
AMERICAN DINOSAUR PALEONTOLOGY

When the term dinosaur was coined in 1842 by the British comparative anatomist Richard Owen (1804–1892), he had only the isolated bones and teeth of three genera of animals—*Iguanodon, Megalosaurus,* and *Hylaeosaurus*—to work with. *Iguanodon* and *Megalosaurus* had been discovered in England in the early 1820s, while *Hylaeosaurus* was found a decade later. There were no complete skeletons available to Owen, nor even a substantial portion of a skeleton, so his reconstruction of what dinosaurs looked like was somewhat speculative. He envisaged them as giant four-footed animals with limbs tucked in under their bodies like the large land mammals of today.[1]

Perhaps the most famous of the bones at Owen's disposal were those of *Iguanodon* from the Lower Cretaceous Wealden sediments of Sussex, first described by Dr. Gideon Mantell (1790–1852) in 1822. Originally misidentified as rhinoceros and hippopotamus bones, Mantell's bones and teeth turned out to be the remains of a giant extinct plant-eating reptile, which he named *Iguanodon* ("iguana tooth") because of the similarity of the larger fossil teeth to those of the modern herbivorous lizard. It had sharp thumb spikes on its hands, although in Mantell's original reconstruction the thumb spike was drawn as a nasal horn, like the horn of a modern rhinoceros.

The other early spectacular find that Owen relied on was the bones of a giant meat-eating reptile from Upper Jurassic deposits near Oxford University. William Buckland (1784–1856), Oxford's first professor of geology, bestowed the name *Megalosaurus* ("large lizard") on this remarkable creature, the first large carnivorous dinosaur to be discovered.

But recent scholarly evidence suggests that a dinosaur discovery in New Jersey predated both Mantell and Buckland's, involving one of America's founding fathers.[2] Dr. Donald Baird, formerly of Princeton University, has suggested that Benjamin Franklin (1706–1790) may have handled a dinosaur bone from New Jersey. Baird has traced the progress of this bone from the Federal period to the present. In 1787, Franklin was the president of the American Philosophical Society (APS) meeting in Philadelphia when he was presented with a large, heavy, darkened bone from Woodbury in Gloucester County, New Jersey. Caspar Wistar, a distinguished Philadelphia physician, speculated that it might be the femur (thigh bone) of a large man. From this meeting of the APS, the bone apparently made its way to the Peale Museum in Philadelphia, the first museum in the United States. The Peale Museum had been founded by Charles Willson Peale (1741–1827), the famous artist who painted portraits of George Washington and many other prominent people of that time. Its eclectic collections included not only works of art and natural curiosities but also the skeleton of a prehistoric mammal, a mastodon that had been discovered at a farm in New York and, at Wistar's urging, excavated, mounted, and displayed by the Peale family. Wistar had arranged for all the APS specimens to be transferred to the Peale Museum.[3] When this museum closed its doors, some of its specimens were transferred to the Academy of Natural Sciences of Philadelphia. Apparently the bone examined by Wistar and Franklin in 1787 was among these objects. Baird's research has identified this specimen, now catalogued as ANSP 15717. The bone is now recognized to be the left metatarsal (foot bone) of a duck-billed dinosaur, although Franklin and Wistar could not have known it as such in their time because Owen's "invention" of the Dinosauria still lay decades in the future.

The first American fossils to be recognized as dinosaur specimens were described in 1856 by Dr. Joseph Leidy (1823–1891) of the University of Pennsylvania and the Academy of Natural Sciences. They were some isolated teeth and fragmentary bones collected during a survey of the Dakota Territory, in what is now Montana, by another Penn professor, Ferdinand Hayden (1829–1887). Among these pieces, Leidy could recognize relatives of *Iguanodon* and *Megalosaurus* in the forms of fossil teeth. These few bits prepared him for a more momentous discovery right in his backyard.

In 1858, another member of the Academy of Natural Sciences, named

William Parker Foulke, was vacationing in Haddonfield in Camden County, New Jersey. Haddonfield was a sleepy Quaker farm town, the scene of several Revolutionary War skirmishes, and the perfect spot to escape from the late summer heat and humidity of Philadelphia. While taking the country air, Foulke became acquainted with one of the largest landholders in Haddonfield, John Hopkins, proprietor of Birdwood Farm and a descendant of the original colonial settlers of Haddonfield.[4] Learning of Foulke's interest in the natural sciences, Hopkins told the Philadelphian of an unusual discovery made some twenty years earlier. While digging marl, a kind of clay that contains fossil seashells, he had come across some very large bones, which he supposed were vertebrae, or backbones. Hearing of the find, visitors came to see the bones, and Hopkins let them carry away some of the vertebrae.

Foulke was very excited by this story. He knew that fossil bones of crocodiles and turtles had been found in the marl pits of southern New Jersey. He asked permission from Mr. Hopkins to dig around the old marl pit where the bones had been found. Hopkins granted him permission "with prompt liberality"[5] and showed Foulke the site of the old marl pit.

The old excavation was by this time slumped in and overgrown, so Foulke assembled a team of marl diggers to help him. The mining of marl was a widespread practice in southern New Jersey, where it was commonly used as fertilizer. Setting to work at the site of the old pit, they dug about ten feet down through thick, dark clay before reaching a layer of fossil seashells containing a cluster of dark, heavy bones. They had hit pay dirt.

The diggers carefully extracted the fragile bones, drawing sketches of their position in the clay and taking measurements of them in case they broke when taken out. The bones were wrapped in cloth and carried up to a cart filled with hay, then transported to the house where Foulke was staying a short distance away.

Foulke realized the importance of the discovery, and he decided to bring in professional help, always a wise decision when dealing with fossil bones. He contacted Joseph Leidy at the Academy of Natural Sciences, who brought with him an expert in fossil seashells, Isaac Lea (1792–1886). Together, this team mapped the area, identified many of the shells found with the bones, and determined that the bones were scientifically valuable and that the digging should continue. The excavation was expanded, and

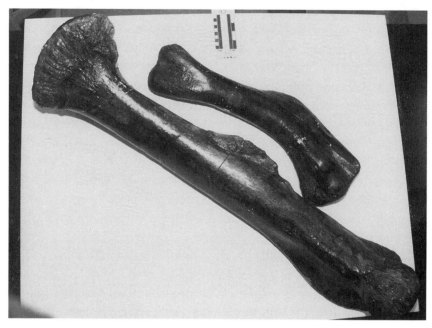

FIGURE 4.1 Humerus and femur of *Hadrosaurus foulkii*. Woodbury Formation, Haddonfield, Camden County, N.J. Casts of type specimens in the collection of the Academy of Natural Sciences of Philadelphia.

the workers dug marl throughout the month of October, 1858, sieving the clay and finding bits of bone and teeth. But no more large bones appeared, and the digging stopped.

In December 1858, William Foulke and Joseph Leidy reported on their work at a meeting of the members of the Academy of Natural Sciences in Philadelphia.[6] Dr. Leidy, in his formal description of the bones, identified them as belonging to a "huge herbivorous saurian" related to the *Iguanodon* of England. He went on to name this animal *Hadrosaurus foulkii*, which in Latin means "Foulke's bulky lizard."

The most important thing about this specimen was that it was the most complete dinosaur skeleton known at this time from anywhere in the world. The bones Owen studied in England were not enough to make a skeleton, and Leidy only had a few teeth from the American West, but in the specimen from Haddonfield there were some forty-nine bones and teeth. This allowed Leidy to make some interesting conclusions about *Hadrosaurus foulkii*. Dr. Leidy demonstrated that the arm bones of *Hadrosaurus* were much shorter than the leg bones of the animal (see figure 4.1). From this, he concluded that this dinosaur was able to rear up on its hind legs and stand straight up like a kangaroo. This was a new way to look at dinosaurs. Previously, Richard Owen's reconstructed dinosaurs for the Crystal Palace exhibition in London were presented as four-legged animals, like giant reptilian elephants. His models, made by the English artist Benjamin Waterhouse Hawkins, were the accepted way to picture dinosaurs until Leidy's description of *Hadrosaurus foulkii*. Today we are used to thinking of some dinosaurs as going about standing on their hind legs (figure 4.2), but it was a very new idea in 1858.

While Owen and Hawkins had produced lifelike, sculpted models of dinosaurs in England, there were no skeletons of dinosaurs mounted in museum displays anywhere in the world. Here, too, *Hadrosaurus foulkii* was the first. Waterhouse Hawkins came to America in 1858 to lecture on natural history and the arts. He took up residence in Philadelphia and convinced the Academy of Natural Sciences, and especially Dr. Leidy, to let him study the bones of *Hadrosaurus.* In November of 1868, he presented the academy with a mounted skeleton made from casts he had made of the bones of *Hadrosaurus,* filling in the parts that were missing (like the skull, which was never found with the rest of the bones).

Thus, Foulke's bulky lizard became the first mounted dinosaur skeleton displayed anywhere in the world—another scientific first for New Jersey's dinosaur! Hawkins made skeletal casts of *Hadrosaurus* for Princeton University, the Smithsonian Institution, and the Royal Scottish Museum in Edinburgh (the first dinosaur skeleton to be displayed in Europe). Hawkins went on to New York, where he was involved in an effort to start a museum of paleontology in Central Park. He built two new skeletal mounts of *Hadrosaurus* for his Paleozoic Museum, as well as other reconstructions of prehistoric animals (figure 4.3). Unfortunately, his plans ran afoul of the infamous Boss Tweed and his Tammany Hall political machine, and so Tweed sent a gang of hooligans to Hawkins's studio. In the dead of night they entered his workshop, broke up all his models and casts, and dumped the pieces in a nearby lake. Disheartened, Hawkins retreated to Princeton University, where he spent several years painting a series of murals depicting scenes of pre-

FIGURE 4.2 Skeleton of *Hadrosaurus foulkii,* showing outline of body and the bones that were recovered by William Parker Foulke and Joseph Leidy in 1858. New Jersey State Museum.

historic life.[7] These murals can still be seen today in the Department of Geology and Geophysics in Guyot Hall on the campus of Princeton University.

There is another reason the Haddonfield discovery was significant. *Hadrosaurus* became an important part of the debates over evolution, the idea that species of animals and plants change over long periods of time. Other anatomists noticed that the New Jersey dinosaur skeleton showed some curious resemblances to the skeletons of modern birds. This led Thomas Henry Huxley (1825–1895), famed defender of Charles Darwin's theory of evolution, to argue that birds had descended from dinosaurs, an idea that many modern paleontologists still endorse. He used the evidence from the skeleton of *Hadrosaurus foulkii* to support his argument.[8]

What about the missing parts of *Hadrosaurus?* Of the vertebrae and other bones that were taken away by curiosity seekers in the original discovery twenty years before Foulke and Leidy dug it up again, local legend says that some of these bones were used as doorstops or window jams. If these

Figure 4.3 Benjamin Waterhouse Hawkins's studio in Central Park, New York. The large skeleton in the center is Hawkins's mount of *Hadrosaurus foulkii;* to the left of the picture the steel frame with leg bones attached is the beginning of Hawkins's reconstruction of *Dryptosaurus aquilunguis*. Flightless bird skeletons can be seen in the background. This is the studio that was destroyed by Tammany Hall hooligans. Courtesy of the Ewell Sale Stewart Library, Academy of Natural Sciences of Philadelphia.

missing pieces had been recovered, the skeleton of *Hadrosaurus* would have been even more complete and may have revealed other interesting features. The greatest loss is the skull, which may never have been buried with the rest of the skeleton. Local residents tell of an attempt by the Academy of Natural Sciences to reexcavate the original site in Haddonfield in the 1920s in an attempt to find the skull. This was apparently unsuccessful, and subsequent investigations at the locality have failed to produce anything more of *Hadrosaurus foulkii.*

Leidy went on to summarize all that was known of dinosaurs and other Cretaceous reptiles in North America in an 1865 publication,[9] including a number of specimens found in New Jersey's marl pits. But soon the field became filled with younger men. Leidy's most promising protégé was a young man named Edward Drinker Cope (1840–1897), the brilliant and quick-tempered scion of a wealthy Philadelphia Quaker family. Studying with Leidy first at the Academy of Natural Sciences and then at the University of Pennsylvania, Cope displayed a precocious talent in anatomy and paleontology. After a year at the Smithsonian Institution, Cope left for Europe shortly after the Civil War broke out. During his tour of European universities and museums, Cope's path crossed the travels of another young American abroad to study paleontology, one Othniel Charles Marsh (1831–1899).

Marsh had important connections of his own. Originally born into rural poverty in the farmland of western New York, Marsh showed an early interest in geology and paleontology. Marsh's uncle was the wealthy mercantilist and philanthropist George Peabody; his rich uncle paid for Marsh's education at Yale University, then sent him on the grand tour to study at the most prestigious centers of learning in Europe. When Marsh returned to America, Peabody endowed Yale with a museum of natural history, where his nephew could pursue his interests.

Cope and Marsh shared an interest in paleontology and so became friendly colleagues. Around the end of the Civil War, both men returned from Europe to begin their careers as serious scientific researchers. Cope obtained a position teaching at Haverford College outside Philadelphia, while Marsh returned to Yale to organize the Peabody Museum of Natural History. They corresponded about their various interests and discoveries and in fact named new species of fossils after each other.

Cope, however, found academic life too restrictive and he soon quit his post at Haverford. Not in need of money, he could afford to pursue his own interests independently. Cope had friends and relatives in southern New Jersey, and so he established a residence in Haddonfield in order to be closer to the marl pits and their fossils. Cope moved his new family to 242 King's Highway in 1868. The Cope house was a Gothic-style mansion situated on a large lot in the center of town; the property included its own water system, a windmill, servants' quarters, and a barn.[10] It was undoubtedly one of the finest homes in Haddonfield. This was Cope's principal residence for eight years, a span of time which his biographer Henry Fairfield Osborn called Cope's most productive period.[11]

By 1866, when Cope was only recently returned from his long European sojourn, he had already established a network of operatives among the marl diggers of southern New Jersey. He would periodically travel out from Philadelphia to make a circuit of mining operations and pick up specimens from the workmen. He obtained numerous vertebrate fossils in this way, including specimens of turtles, crocodiles, mosasaurs, and his first dinosaur bone, a hadrosaur femur (see figure 4.4). One set of bones excited him greatly, a partial skeleton obtained from Alfred Voorhees of the West Jersey Marl Company, near Barnsboro, Gloucester County, New Jersey (see figure 4.5). These fossils included a very large claw, a section of lower jaw with incipient replacement teeth, several sharply pointed, laterally compressed teeth with serrated edges, plus leg and foot bones. Here, Cope knew, he had the carnivorous relative of the English *Megalosaurus* and the *Deinodon* of western America.

Cope delivered a preliminary description of the animal's remains before the Academy of Natural Sciences in August 1866.[12] It is clear from his initial report that he regarded it as a very active predator; he speaks of its birdlike great claw, which he regarded as "prehensile" and "contractile." He also thought that it was bipedal, by analogy with Leidy's *Hadrosaurus,* and that it attacked in a leaping fashion, using its hind feet for inflicting "fatal wounds." He named this dinosaur *Laelaps aquilunguis,* meaning "eagle-clawed terrible leaper."

Cope's reports on the discoveries in the New Jersey marl pits attracted the attention of his colleague O. C. Marsh up in New Haven. Shortly after Cope moved to Haddonfield, Marsh contacted him and arranged for a

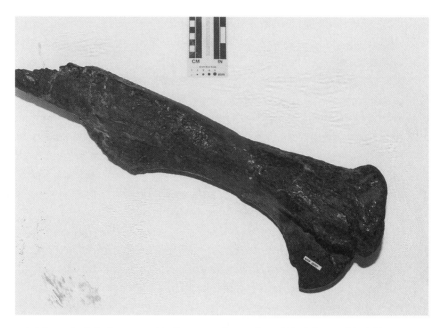

FᴵɢᴜʀᴇCFIGURE 4.4 Cope's first dinosaur: *Hadrosaurus* femur from West Jersey Marl Company pit near Barnsboro, Gloucester County, N.J. Collections of the Academy of Natural Sciences of Philadelphia.

tour of the marl pits. Cope graciously hosted Marsh in Haddonfield in the spring of 1868, and the two traveled out to the marl mines by horse-drawn coach. But while Cope was showing Marsh around his localities, Marsh was familiarizing himself with more than just the geology and the fossils; he became acquainted with the marl diggers as well. For after 1868, the bones from the marl pits that had been going to Philadelphia's Academy by way of Haddonfield seemed to make a right-hand turn and wound up in New Haven at the Peabody Museum instead. Marsh had suborned Cope's network of marl-pit workers, most notably Alfred Voorhees, the deliverer of *Laelaps*.

As it slowly became apparent that Marsh was skimming off the prize finds from the marl pits, Cope's distrust of his colleague from Yale grew. He became increasingly suspicious of Marsh's motives, and the feeling of professional cooperation was replaced by heightened hostility. When Marsh began prospecting the rich fossil fields of western Kansas in 1870, it was not

FIGURE 4.5 Operations at West Jersey Marl Company Pit near Barnsboro, N.J., September 1888. Courtesy of Gloucester County Historical Society, Woodbury, N.J.

long before Cope took an interest in the same area, perhaps in retaliation for Marsh's claim-jumping in New Jersey. Cope obtained a large skeleton of a marine reptile called a plesiosaur from the Cretaceous marine beds in Kansas, by way of Theophilius H. Turner, a U.S. Army surgeon at Fort Wallace. Cope named his new fossil *Elasmosaurus* and had the bones mounted as a skeleton in the Academy of Natural Sciences' public museum. With great fanfare, Cope invited the leading scientists of the day to the great unveiling of his latest Mesozoic monster. At the appointed hour before the assembled guests the cover was drawn off the skeleton, only to reveal a horrendous mistake. As Marsh pointed out to all the luminaries present, Cope had mounted the head on the end of the tail of the animal!

Cope could never forgive this public humiliation, and an intense rivalry between the two men began which became known subsequently as the "bone wars." Each man tried to outdo the other in the discovery, description, and naming of new fossil organisms. This led to such subterfuges as spying on each other's field operations, claim-jumping, secret codes, and double agents. The two rivals eventually wound up accusing each other in the public press of unseemly behavior. While many historians of science have pointed to the incident of the plesiosaur head as the starting point of the Cope-Marsh feud, the bone wars actually got their start in the marl pits of southern New Jersey.

Laelaps was one of the casualties. In another one of the many salvos fired in the bone wars, Marsh pointed out in 1877 that the generic name *Laelaps* was already being used for another animal (a spider), and, by the conventions of biological nomenclature, it could not be applied to another organism, so he renamed Cope's pride and joy *Dryptosaurus* ("wounding reptile"). This must have really made Cope angry, but there was not much he could do about it except to ignore Marsh's new (and valid) name. Cope continued to apply the name *Laelaps* to carnivorous dinosaur teeth he found in Montana.

After the discovery of the great dinosaur-bearing beds of the Morrison Formation in Colorado and Wyoming in 1877, the real action in vertebrate paleontology moved to the American West. Cope and Marsh expended their personal fortunes on obtaining fossil bones. Cope's declining financial situation was exacerbated by a series of bad investments in silver mines in New Mexico. Cope eventually had to accept employment at the University of Pennsylvania, and he sold off part of his fossil collection to the new American Museum of Natural History in New York. Marsh's fossils (including the New Jersey specimens) became the core of the Yale Peabody Museum's collections. In the meantime, the marl-mining industry in New Jersey declined and nearly disappeared.

In late March 1897, Edward Drinker Cope, seriously ill and only weeks away from death, was visited by a young artist from New York, Charles R. Knight.[13] Among the very last of his many contributions to paleontology, Cope instructed Knight on the pose and attitude of several dinosaur paintings the young artist was working on for *Century* magazine. Included in this group was a lively reconstruction of *Dryptosaurus* (*Laelaps*), Cope's first great dinosaurian discovery and thus an animal close to his heart.

Many people point to John Ostrom's discovery and description of *Deinonychus* in the 1960s as a turning point that changed our way of thinking about dinosaurs, and certainly this was an important step in revising the

FIGURE 4.6 Charles R. Knight's painting of "Leaping Laelaps" (now known as *Dryptosaurus*) done under the direction of E. D. Cope. Negative/Transparency # 335199, courtesy of the Department of Library Services, American Museum of Natural History, New York, N.Y.

old image of dinosaurs as essentially all slow, stupid animals. This small, heavily clawed predator was a very active animal, obviously built for speed and agility in its attack.[14] *Deinonychus* is often depicted as leaping at its prey (usually the ornithopod *Tenontosaurus*), feet first with deadly claw extended.

This is precisely the pose dictated by Cope to Knight in those final days of the great paleontologist's life. The depiction of "Leaping Laelaps" done by Knight in 1897 (figure 4.6) presages by six decades the more modern image of the dinosaurs as active, alert animals, and Cope's 1866 description of "*Laelaps*" anticipates by a full century the revolution in the image of the dinosaur that has come to public fruition in Steven Spielberg's movie *Jurassic Park. Laelaps* was Cope's first and last testament to dinosaur paleontology, and a telling tribute to his abilities as a scientist.

As for *Hadrosaurus foulkii,* its scientific and historic importance has recently been recognized in a series of formal ceremonies. The site of the discovery was first indicated by a stone marker erected in 1984 by a local boy scout troop. In 1991, *Hadrosaurus foulkii* was officially declared the State Dinosaur of New Jersey. This honor was the result of years of effort by teacher Joyce Berry and her fourth-grade classes at Strawbridge Elementary School in Haddon Township, New Jersey. In 1995, the National Park Service designated the site of *Hadrosaurus foulkii*'s discovery as a National Historic Landmark. This status is noted on a plaque near the original excavation, at the end of Maple Street in Haddonfield.

CHAPTER FIVE

THE EARLIEST DINOSAURS

In the Triassic Period, the first part of the age of dinosaurs, most of the land-masses of the earth were united in a great supercontinent called Pangaea. South America's great bulge of Brazil was tucked into the bight of west Africa, Eurasia linked to Canada by way of western Europe and the mar-itime provinces, and the east coast of the United States was connected to North Africa. New Jersey was next to Morocco. Pangaea's interior regions (such as New Jersey) were subjected to the severe extremes of continental climate. We see such extremes on a smaller scale on today's continents; the inland regions tend to have drier weather and greater daily and seasonal temperature fluctuations than the coasts, where the oceans act as a buffer on climate. The harsh, arid world of Pangaea was the environment that gave rise to the first dinosaurs. The earliest dinosaurs now known are from the Ischigualasto Formation of Argentina, from the late Middle Triassic approxi-mately 230 million years ago. These first dinosaurs are mostly small, lightly built, agile meat-eaters, including *Eoraptor, Herrarasaurus,* and *Stauriko-saurus.* They were not immediately abundant; their remains are not as com-mon as the fossil skeletons of mammal-like reptiles or specimens of their archosaur cousins.[1]

In New Jersey, the early dinosaurs are represented by the foot impres-sions they left in the drying red Pangaean mud. Today these fossil footprints are found in the redbeds of the Newark Basin, a wide swath of sediments stretching across the state from the Hudson River to the Delaware River between Trenton and Milford (see map in figure 5.1). New Jersey's land-scape was a lot more rugged at the end of the Triassic Period than it is today; high young mountains alternated with deep valleys formed by faults,

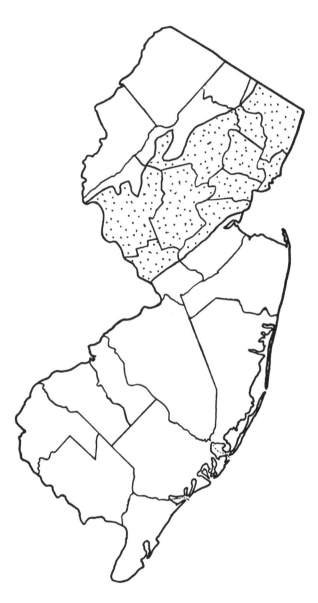

FIGURE 5.1 Geologic map of Newark Supergroup Triassic and Jurassic rock formations in New Jersey (stippled area).

FIGURE 5.2 Newark Supergroup Triassic and Jurassic rock formations in New Jersey (chart).

PERIOD	FORMATION	FOSSILS
Jurassic	Boonton	fish, early crocodile tracks
	Hook Mtn.	(basalt; no fossils)
	Towaco	dinosaur tracks, fish, insects, plants
	Preakness	(basalt; no fossils)
	Feltville	dinosaur and crocodile tracks, fish
	Orange Mtn.	(basalt; no fossils)
Triassic	Passaic	dinosaur tracks, thecodonts, fish
	Lockatong	fish, phytosaurs, amphibians, insects
	Stockton	large amphibians, phytosaurs, plants

cracks in the earth that dropped down blocks of crust to form long linear troughs. During the brief rainy season, flashfloods filled intermittent streams with sediments eroded from the high mountains on either side of the trough. The swollen streams dropped their load of mud and silt in the lower areas of the trough. During the dry seasons, as the muds and silts dried out and cracked in the sun, early dinosaurs walked over the baking mud, leaving their footprints on the glazed, mud-cracked surfaces. Eventually these surfaces were buried by the next flashflood's load of sediments, preserving the footprints and other surface features. This process was repeated for millions of years, burying the hardened muds deeper as the fault block trough continued to drop down and more mud piled up.

The flood waters flowed into large lakes called playas, which occupied the central, most depressed areas of the troughs. During the rainy seasons these bodies of water grew in size, at times covering most of the bottom of the long rift valley; during protracted dry seasons, the playa lakes shrank and became salty, like Great Salt Lake in Utah today. Brine shrimp prospered in their salty waters, and at times large populations of primitive bony fish died off in massive fish kills; their bodies floated to the bottom, to be buried and fossilized. Today, certain thin layers of the lakebed sediments, now a hard, dark rock called argillite, yield fossils of coelacanths, semionotids, and other prehistoric fish. One such layer was encountered during

it was about a foot long at most. *Hypsognathus* is known from both skeletal remains and footprints in the Newark Basin Triassic deposits. One specimen, a skull, was found in a brownstone wall in Passaic, Passaic County, New Jersey (figure 5.5).

Another small reptile of Late Triassic New Jersey was the "armadillo reptile," *Stegomus arcuatus.* This armored animal had a series of bony plates encasing its body, as well as a longish snout like a modern armadillo, and one might imagine a similar lifestyle of poking about for insects and grubs as it waddled over the mudflats.[3] Its remains have been found at Neshanic Station in Hillsborough Township, Somerset County, and in Alexandria, in Hunterdon County.

One of the stranger small reptiles that lived around the ancient lakes was a creature scientists call *Tanytrachelos.* This animal had a lizardlike body with sprawling legs and squat torso, but its small head was perched at the end of an elongated thin neck. It probably was an aquatic animal, and it may have fed on water bugs or small fish. Its skeletal remains are usually found in the lakebed deposits.

While all the animals described above are known from fossil bones, there are indications of other creatures haunting the lakesides and floodplains. The evidence for these animals is in the form of footprints. We can deduce something about these creatures even though their tracks are the only evidence we have. For instance, three-toed tracks that occur only as footprints (without associated forefeet impressions) reveal the presence of creatures walking on their hind legs, the stance adopted by the early carnivorous dinosaurs. The three-toed clawed footprint impressions match the known structure of predatory dinosaur feet. Earlier in the Late Triassic, there do not appear to have been many dinosaurs present in the Newark Basin; the few dinosaur footprints found in the Stockton Formation and the Lockatong argillite are generally small (3 to 4 inches) and rare. Larger footprints belonging to animals called rauisuchids are given the name *Brachycheirotherium.* These animals seem to have been the dominant large predators in the Triassic, while the earliest dinosaurs were smaller. In contrast to the dinosaur footprints, *Brachycheirotherium* trackways are sets of hand and foot impressions that show four fingers and five toes. Similar but slightly smaller tracks are called *Cheirotherium.*

As the Triassic Period drew to a close, conditions in the Newark Basin

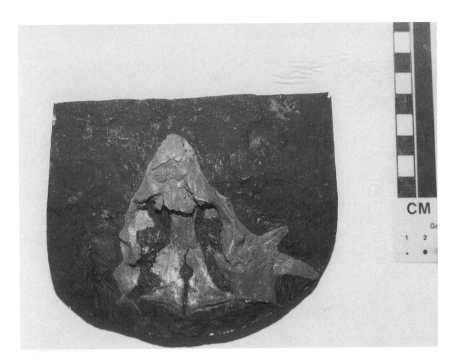

FIGURE 5.5 (Top) Skull of *Hypsognathus fenneri* from Passaic, Passaic County, N.J. (cast). New Jersey State Museum. (Bottom) Model of *Hypsognathus fenneri*. New Jersey State Museum.

FIGURE 5.6 Model of early bipedal dinosaur. This is the kind of small theropod thought to have made the dinosaur footprints called *Grallator.* Courtesy of William Selden, Rutgers Geology Museum, New Brunswick, N.J.

became more arid and desertlike. Following the widespread deposition of the Lockatong lakebed sediments, redbeds consisting of more oxidized muds, silts, and sands were laid down by floods across the Newark Basin rift valley. Along the steeper faulted walls of the trough, coarser sediment composed of cobbles and gravels were deposited as the power of flashfloods moved larger pieces of rocks down from the surrounding mountains and into the margins of the rift valley. The rainy spells were followed by long periods of dryness and desiccation; geologists can tell that it was dry because of the presence of certain kinds of soil deposits in the redbeds, deposits called caliche. Today, calcareous caliche deposits can be seen forming in the dry soils of the arid southwestern United States. The thick sequence of Late Triassic redbeds (with some smaller lakebed deposits at various levels) is called the Passaic Formation. As one goes upward through the stack of Passaic rocks from the earliest beds to the most recent, dinosaur footprints become more frequent and also appear to get larger. The smaller

FIGURE 5.7 *Anchisauripus* and *Grallator* footprints. The larger footprint is the natural mold (or negative) of *Anchisauripus;* to the immediate left of this (below the letter "H") are several smaller and fainter *Grallator* footprints heading in the opposite direction. Courtesy of Rutgers Geology Museum, New Brunswick, N.J.

three-toed form called *Grallator* is still present, but in addition there is a larger three-toed footprint called *Anchisauripus* (figures 5.7 and 5.8).

Other kinds of footprints indicate the presence of sphenodontians, primitive relatives of the lizards. Small four-footed prints called *Rhynchosauroides* are interpreted as sphenodontian tracks. There is one surviving representative of this group, the endangered tuatara of New Zealand, a true living fossil.

In addition to *Cheirotherium* tracks and phytosaur footprints (called *Apatopus*), several new sorts of trackways begin to appear in the Passaic Formation redbeds. Three-toed footprints associated with four-toed handprints have been given the name *Atreipus;* the animal that made these impressions may represent a very early form of the ornithischian dinosaurs. Another kind of small four-toed footprint is called *Batrachopus;* these trackways are thought to have been made by early small crocodiles.

In the very last beds of the Passaic Formation, approximately a meter

FIGURE 5.8 Two *Anchisauripus* footprints headed in opposite directions. These prints are natural casts (or positives) formed by infilling sediment. The fleshy toe pads and claw impressions are clearly visible, especially in the track on the left. Courtesy of Rutgers Geology Museum, New Brunswick, N.J.

(about a yard) below the first of the Watchung Mountain lava flows, there is a remarkable change in the nature of the fossils. All the fossil clues point to a sudden turnover in the plants and animals that lived in this region. What is even more remarkable is that this change is reflected in the fossil record all over the world. Dinosaur footprints become more abundant, and larger forms are more common. There is also a sudden change in the fossil pollen.

Pollen from plants is very common in the fossil record, at least in deposits laid down after the advent of land plants. In prehistoric times, as it does today, pollen drifted in the atmosphere in tremendous amounts. While microscopic, pollen is nonetheless very durable and is easily buried and preserved in vast quantities. Individual grains of pollen have distinctive forms according to what species of plant produced them; different species produce different pollen forms. The changes in the pollen grains and of the plants that produced them over time give geologists an important indicator of specific spans of prehistoric time. In the rocks of the uppermost part of the

Passaic Formation, the sudden change in pollen fossils is the change from pollen forms of latest Triassic age to pollen forms of earliest Jurassic age.[4] This switchover is called the Triassic-Jurassic boundary. It marks a major turnover in the fauna and flora of the Mesozoic Era, a mass extinction event that terminated one period of geologic time and started another. Typical Triassic animals such as the thecodont-grade archosaurs disappeared, and dinosaurs became the dominant animals on land.

Very shortly (geologically speaking) after this changeover, the supercontinent of Pangaea began to break up. Some of the best evidence for the beginning of this crustal fragmentation is in the form of the several thick lava flows that are found in the upper part of the Newark Basin rocks, not only here in New Jersey, but also in several other fault block basins in Pennsylvania, the Connecticut River Valley, and Nova Scotia. This widespread outpouring of lava and the associated underground igneous intrusions indicated a stretching and cracking of the supercontinental crust that would eventually result in the formation of the Atlantic Ocean. Today, the early Jurassic lava flows form the basalt of the several Watchung Mountains; the cooled and hardened igneous intrusions underlie the Hunterdon Hills and form the Palisades.

In the sedimentary rocks of the Feltville and Towaco Formations laid down in between the lava flows, we see a continuation of the tendency for dinosaurian size increase in the footprint fossils. In the early Jurassic rocks, large foot impressions of the type called *Eubrontes* (figure 5.11) show that the first of the large meat-eating dinosaurs roamed across New Jersey at this time, stalking the other animals in the Newark Basin. While no bones of these larger carnivorous dinosaurs have been found in New Jersey, the kind of animal that might have made similar footprints has been dug up in the American Southwest. This is the dinosaur called *Dilophosaurus,* a twenty-foot long predator of early Jurassic time whose skeletons have been excavated in Arizona. Contrary to fanciful movie reconstructions of this animal, there is no evidence that it could spit poison or that it had a retractable collar. It did, however, have a pair of thin bony crests along the top of its head and snout (hence its genus name; *Dilophosaurus* means "two-ridged reptile"). The precise function of these crests is as yet unknown; it is speculated that they may have served some social function, such as attracting mates or establishing status during competition between these dinosaurs.

FIGURE 5.9 *Grallator* footprint from a small dinosaur. Note toe pads and claw marks. From West Patterson, Bergen County, N.J., uppermost Passaic Formation, Early Jurassic. New Jersey State Museum.

A good place to see dinosaur footprints in New Jersey is at the Walter T. Kidde Memorial Dinosaur Footprint Quarry in Roseland, Essex County, New Jersey. Here, one can find small *Grallator* footprints, intermediate-sized *Anchisauripus* footprints, and large *Eubrontes* footprints in the Towaco Formation redbeds. Some authorities (such as Paul Olsen of Columbia University) believe that these footprint assemblages represent growth stages of the same animal, rather than several different genera of dinosaurs.[5] Certainly all three footprint types are of the same general form with three clawed toes and the same basic shape. In the past some paleontologists have further subdivided these three subgenera into numerous different species. This "splitting" is even more suspect, because footprint form can be very different, depending on variables like grain size of the sediment, wetness of the sediment, and gait of the animal. So the same kind of dinosaur may have made different footprints, depending on whether it was running or walking on wet or dry mud, silt, or sand.

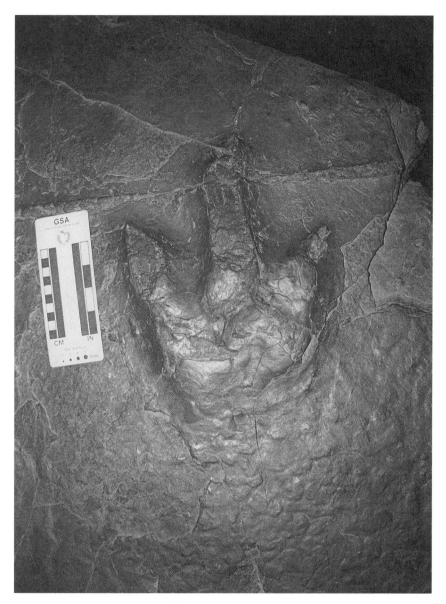

FIGURE 5.10 *Anchisauripus* footprint from an intermediate-sized dinosaur. Natural cast on mudcracked surface, from Lincoln Park, Morris County, N.J. Towaco Formation, Early Jurassic. New Jersey State Museum.

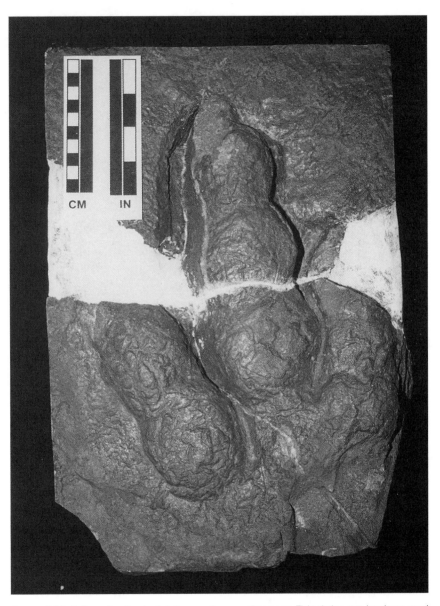

FIGURE 5.11 *Eubrontes* footprint from a larger dinosaur. This slab was broken and repaired with plaster. From West Patterson, Bergen County, N.J., uppermost Passaic Formation, Early Jurassic. Donated by Chris Laskowich and repaired by Donald Baird. New Jersey State Museum.

Another footprint form found at the Roseland quarry suggests that advanced synapsids were present in the Newark Basin. Paul Olsen[6] believes that small five-toed hand and foot impressions, first found by Larry Felder in 1977, represent the tracks of therapsids, synapsid reptiles that were very close to the ancestry of true mammals. It is a curious feature of the fossil record of the Mesozoic that while the first mammals and early dinosaurs appeared quite closely in time, dinosaurs went on to become the large prominent animals of the Mesozoic world while mammals were relegated to a distinctly subordinate role as small, probably nocturnal, insect eaters during the Jurassic and Cretaceous Periods.

At the top of the Newark Basin sediment stack, after the last lava flow (the Hook Mountain Basalt), wetter conditions seem to have briefly prevailed during deposition of the Boonton Formation. Lakebed muds accumulated, and a massive fish-kill was preserved in these sediments. At Boonton in Morris County, New Jersey, these fish were preserved as darkened carbonized films that display the distinct body outlines of the animals. Shiny diamond-shaped scales are preserved in exquisite detail (figure 5.12), as are details of the heads, fins, and tails. They represent primitive bony fishes called *Semionotus* and *Redfieldia*. *Semionotus* is related to the modern gar fish, while *Redfieldia* is closer to the paddlefish, a bizarre-looking fish inhabiting the Mississippi and Missouri rivers today.

The sedimentary rocks of the Newark Basin contain the record of the early dinosaurs' rise to become the rulers of Mesozoic earth. Footprint fossils, initially small and uncommon, demonstrate the tendency toward larger size that would characterize subsequent dinosaur evolution. The mass extinction event at the end of the Triassic Period seems to have opened the way for the dinosaurs to become the dominant land animals of the Mesozoic Era. Most of the previously important non-dinosaurian Triassic animals of the lower part of the Newark Supergroup, such as the metoposaurs, phytosaurs, *Hypsognathus, Stegomus,* and rauisuchids, declined and disappeared by the end of the deposition of the Passaic Formation rocks. These extinct animals were replaced by a fauna of dinosaurs and protosuchid crocodiles that became more common in the early Jurassic part of the Newark Supergroup beds.[7]

It has been speculated that perhaps the impact of a large asteroid or comet caused the extinctions at the end of the Triassic Period. Indeed, there

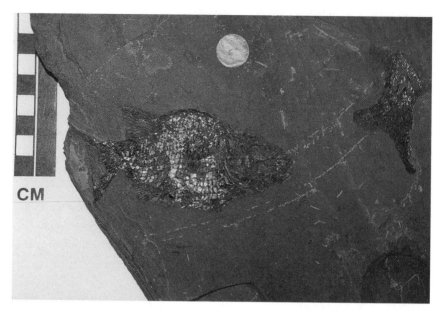

FIGURE 5.12 Carbonized fish fossil from Boonton, Morris County, N.J. This *Semionotus* specimen is typical of the fish-kill layers in the lakebed sediments of the Newark Basin. Boonton Formation, Early Jurassic. New Jersey State Museum.

is a large crater in Canada called Manacouagan that is about the right age to coincide with the mass extinction event at the Triassic-Jurassic boundary. It is curious, however, that while the dinosaurs survived and subsequently thrived after this event, a similar impact scenario has gained acceptance as an explanation for their extinction at the end of the Cretaceous Period (as we shall see in Chapter 9). Perhaps some other factor affected their survival, such as increased size, or the breakup of Pangaea.

For the early dinosaurs were Pangaean animals. We find the same kind of small three-toed bipedal footprints all over the Late Triassic and Early Jurassic rocks of the old supercontinent, in western North America, Europe, Iran, and South Africa. It is tempting to imagine that the early dinosaurs' ability to rear back on their two hind legs and go about in a more or less upright fashion gave them some important advantages over their four-footed reptilian cousins. Bipedality may have helped the dinosaurs' widespread dispersal over the broken, rugged terrain of Pangaea. It also allowed them to stand up and look around for food, water, or approaching predators, and freed the forelimbs for grasping. These adaptations, forged in the harsh Pangaean environment, served the dinosaurs well during their rise to become the largest, most diverse animals on earth for the next 150 million years.

CHAPTER SIX

HEYDAY OF THE DINOSAURS

In 1870, Joseph Leidy received a single broken vertebra collected by Ferdinand Hayden from near Canyon City, Colorado. Leidy, the namer of *Hadrosaurus,* described this new bone before the Academy of Natural Sciences and recognized it as a new kind of extinct reptile, tentatively assigning it the name *Antrodemus.*[1] This was the first of what was to be a treasure trove of dinosaur bones from the Late Jurassic deposits in the American West. By 1877, when several major discoveries focused interest on fossil beds at Morrison and Canyon City in Colorado and then at Como Bluff in Wyoming, Cope and Marsh had already abandoned the New Jersey marl pits for the richer fossil fields in Kansas, Nebraska, New Mexico, and Montana. The greatest finds were to come out of the hogbacks and badlands of the Morrison Formation, a widespread deposit of distinctively banded red, green, and cream-colored rocks laid down from about 150 to 144 million years ago. At these sites, Cope's and Marsh's field crews tried to outdo each other in excavating new gigantic dinosaurs of Jurassic age. Out of the river channel and floodplain deposits these pioneer paleontologists would unearth many of the dinosaurs most familiar to us today: the Jurassic pantheon of *Apatosaurus, Camarasaurus, Diplodocus, Stegosaurus,* and *Allosaurus.*

The Late Jurassic was the heyday of the giant dinosaurs. By Late Jurassic time dinosaurs had evolved into a wide diversity of different types, some of them enormous. Our knowledge of the Morrison giants has grown since the days of Cope and Marsh. In 1900, Elmer Riggs discovered a partial skeleton, which he named *Brachiosaurus,* in Morrison beds near Grand Junction, Colorado. This genus turned out to be the most massive of all the great sauropod dinosaurs. Since then, even larger brachiosaurs have been found in western Colorado; these have tentatively been named "Supersaurus" and "Ultrasaurus." Since both of these specimens are only partial skeletons, their actual size can only be estimated, but informed guesses place their length at 98 feet and weight at close to 130 tons.[2] Even more recently, a truly enormous skeleton has been excavated in northwestern New Mexico; christened *Seismosaurus,* this relative of *Diplodocus* is estimated to have been 150 feet long!

Unfortunately, there are no deposits of Late Jurassic age exposed in New Jersey or anywhere else close by. New Jersey was largely mountainous during that time, and so there was probably net erosion in this region rather than deposition of sediments. The nearest rocks of Late Jurassic age are deeply buried offshore in the Baltimore Canyon trough near the edge of the continental shelf. But because the western dinosaurs could migrate freely across an unbroken continent, and dinosaurs very similar to the Morrison Formation giants have been found in Portugal and Africa, it is likely that Late Jurassic dinosaur faunas in this area were similar to the famous dinosaurs of the American West.

Our best proof for this idea comes from the lower Cretaceous deposits found not very far south of New Jersey. In a belt of sands and clays extending between Washington, D.C., and Baltimore, Maryland, old iron mines and sand pits have yielded a fragmentary dinosaur fauna that offers us a glimpse of the heyday of the dinosaurs in the eastern part of North America. The deposits of the Potomac Group record the erosional leveling of the ancient Appalachian Mountains as the great supercontinent of Pangaea ripped apart at the seams and the baby Atlantic Ocean grew larger. The eroded material carried by the ancient rivers and streams was laid down as a broad apron of sediments deposited in low-lying floodplains and marshes. In some of the old river channel deposits that were cut off from the main current, swampy conditions favored the burial and preservation of fossils. Here abundant plant remains and the isolated bones and teeth of dinosaurs were preserved in a subterranean environment that was also conducive to the precipitation of iron from groundwater. Over 100 million years later, these deposits were mined in the nineteenth century to provide iron for the steel of the Industrial Revolution. During the course of that excavation, the miners found a variety of bones and teeth. Hearing of these discoveries, O. C. Marsh dispatched one of his best field workers, John Bell Hatcher. Hatcher secured a number of specimens from the old mines for the Yale Peabody Museum. Other specimens from the Arundel Formation, as the bog-iron deposits came to be known, were donated to the nearby Smithsonian Institution.

The fossils from the Arundel Formation revealed a dinosaur fauna in transition. There were several indeterminate meat-eating dinosaurs of large and small size, known mostly from a few isolated bones and teeth. The teeth and bones of a sauropod called *Astrodon* are the most common dinosaur remains found in these Maryland fossil beds; in fact, the first sauropod fossils known from North America were teeth of *Astrodon* described in 1859 by Dr. Christopher Johnson.[3] The numerous fragmentary remains indicate several species of smaller sauropods, from 12 to 30 feet in length.[4] Just recently a very large sauropod femur was discovered in the Arundel sediments in a sand pit near Muirkirk, Maryland, demonstrating that larger animals were also present. What is known of *Astrodon* suggests that it is related to the gigantic Morrison brachiosaurs of the west.

But there also is evidence of a new element in the fauna, in the form of dinosaurs like *Priconodon,* an armored nodosaur, and *Tenontosaurus,* a relative of *Iguanodon. Tenontosaurus* was an ornithopod, a group that included the duck-billed dinosaurs of the Late Cretaceous, and as such it was a forerunner of *Hadrosaurus foulkii.* This family of dinosaurs had an advanced jaw construction with rows of teeth that could shear past each other in a grinding motion. This design permitted a kind of chewing which helped with processing the tough fibrous plant food that made up their diet. The Morrison giants, the sauropods and stegosaurs, had much different methods for digesting their plant food. The sauropod mouth was basically a food rake with a set of either peg-shaped or leaf-shaped teeth that merely ripped the vegetation away from the plant. The food was then swallowed whole to be processed in the gut, probably with the assistance of stomach stones or gastroliths. Today, we see animals like birds and alligators swallowing pebbles to help them grind up food in their stomachs. Polished piles of stones are sometimes found associated with large sauropod skeletons,[5] and it is hypothesized that the animals used these gastroliths as an aid in digestion. *Tenontosaurus* and its later relatives represent an alternative way to eat plant food, perhaps as a response to changing vegetation.

For in the same beds that yield the Arundel Formation dinosaur remains, there is evidence of a profound change in the Mesozoic flora. Paleobota-nists working at the Smithsonian Institution have found fossil evidence of some of the earliest flowering plants in the Potomac Group deposits. Through most of the Mesozoic Era, the dominant vegetation had consisted of gymnospermous plants, plants with naked seeds like conifers and cycads. Other important plant groups included the ferns and the horsetail rushes.

But early in the Cretaceous the first leaf impressions, seeds, and pollen of true flowering plants (also known as angiosperms) appeared. They slowly took over the landscape, starting with the unstable environments of river-banks and spreading along coastal areas. The early angiosperms had rela-tively disorganized leaf structure and simple pollen grains. As the early flowering plant fossils are traced upward in successively younger strata, the leaves develop more organized venation and the pollen becomes more complex.

In New Jersey, that change in flora is evident in the beds of the Potomac Group and the overlying younger strata of the Raritan Formation. The Late Cretaceous Raritan fossil flora has a greater diversity of more complex flowering plants, although large conifers like *Metasequoia*, the dawn red-wood, were prominent in the landscape. By this time, however, some of the modern angiosperm groups had appeared, including ancient relatives of the magnolias, the sycamore trees, and the rose family. The landscape began to change as the more advanced flowering plants spread across the earth.

Some authorities think that the change in dinosaur faunas in the Early Cretaceous is linked to the change in vegetation. As plants changed, so did the dominant groups of herbivorous dinosaurs feeding on them. According to this view, the sauropods and the other vegetarian dinosaurs of the Mor-rison fauna gave way to new groups of plant-eaters in the Cretaceous that were better equipped to consume the spreading flowering plants.

At least one dinosaur expert sees it differently, however. According to Robert Bakker,[6] the dinosaurs may have paved the way for the flowering plants by efficiently mowing the undergrowth and creating conditions con-ducive to the growth of angiosperms. Flowering plants can grow more quickly and reproduce more rapidly than gymnosperms, and thus have an advantage in closely cropped areas where animals are intensively grazing.

In Bakker's view, the replacement of the older high-browsing Morrison fauna of giants by the Cretaceous dinosaurs that ate closer to the ground was the reason for the success of the flowering plants. Of course most paleobotanists think that this is putting the cart before the horse; they point out that most plant-eaters adapt to the changes in the vegetation rather than the other way around. Whatever happened, the low-browsers of the Early Cretaceous gave rise to the important dinosaur groups of the later part of the period, and the sauropods and stegosaurs of the Morrison fauna became a much less significant portion of the dinosaur assemblage.

One of the advantages enjoyed by flowering plants is pollination by insects. Very early in their evolution, the pollen grains of flowering plants began to develop complex surfaces, sculpted with hooks and spines and other projections designed to give the pollen grain free transportation. The projections aid the pollen in hitching a ride on the legs or body fur of an insect visiting the flower in which the pollen and nectar are produced. As the insect goes from flower to flower collecting nectar, it also deposits the pollen and fertilizes the plant. The plant can then produce seeds that will be enveloped by the shell of nuts, or the fleshy edible coverings of berries or fruit. The seeds will grow into new plants. Many angiosperms have come to depend on insects to assist their reproduction.

From the Raritan and Magothy Formations of New Jersey, important information about the early phases of this coevolutionary relationship is coming to light. In the same strata that produce the fossil leaves, flowers, and pollen grains from the early Late Cretaceous flowering plants, layers of lignitic clay (figure 6.1) yield abundant pieces of amber, fossilized tree resin from the age of dinosaurs. The amber is a translucent golden preservative that will safeguard whatever is entrapped within it. Insects engulfed in the flowing tree sap 90 million years ago are preserved in exquisite detail; fossil spiders display intact all the minute hairs on their bodies and legs.

New Jersey fossil-bearing amber has produced the oldest known fossil bees: stingless honeybees first found in amber from Kinkora in Burlington County. Because this specimen looks very similar to modern worker bees, scientists conclude that social organization had already developed among these insects. Moreover, many advanced anatomical features (such as the structure of the hind legs) similar to those of modern bees suggest that the Cretaceous bees were among the earliest pollinators of flowering plants.[7]

FIGURE 6.1 Digging for amber in the Raritan Formation, near Sayreville, Middlesex County, N.J. Darker material is rich in lignite, fossil wood from the Cretaceous Period.

New Jersey has also produced some of the oldest known ants, first described from amber pieces found along the shore at Cliffwood Beach near outcrops of the Magothy Formation. Anatomical details of these primitive insects reveal that ants probably arose from wasps.[8] The oldest fossil moth egg was found in deposits of the same age exposed at Gays Head cliffs on the southwest end of Martha's Vineyard in Massachusetts.[9]

Extensive collections of fossiliferous amber have been made by many amateur collectors, notable among which are the specimens collected by Penney Dillon and Ralph Johnson in the Sayre-Fischer pit and adjacent excavations in Middlesex County. Under the microscope the yellow droplets reveal a fantastic fauna of insect forms, many of which are familiar to us today. There are pieces that contain three beetles all in a row, as if one were following another when they were overtaken by the flowing sap. Other pieces show individual beetles colored bright green, displaying the original pigmentation. The delicate venation in the wings of flying insects is pre-

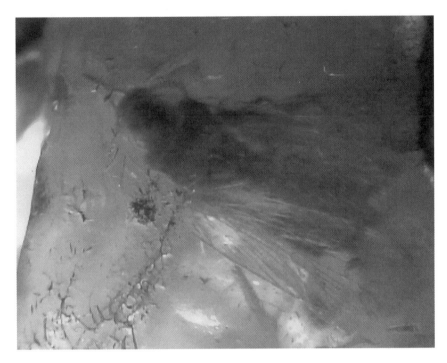

FIGURE 6.2 Cretaceous cockroach in amber. From Raritan Formation, Sayreville, Middlesex County, N.J. Courtesy of Edward Gilmore.

served in remarkable detail. There are small cockroaches (figure 6.2), ants, flies, midges, mites, lacewings, leafhoppers, and several kinds of stingless wasps and bees, all mummified for the ages. The most intricate and complex structures, like the compound eyes of some insects, are amazingly visible. There are even fragments of Cretaceous bird feathers preserved in the amber, although these are very rare. Not only are the New Jersey amber deposits extremely productive and informative, they are also among the very oldest fossiliferous amber beds known; the amber mines of the Dominican Republic made famous by the movie *Jurassic Park* are far too young at 25 million years old to have yielded bugs from the age of dinosaurs. For genuine Mesozoic amber containing prehistoric insects, New Jersey is one of the best spots in the world; only the insect-containing amber from Lebanon is slightly older, and the small amount of amber found in beds of Jurassic age has not yet produced any fossil insects.

Insects and bird feathers are not the only fossils found in amber. A few small flowers have been found in Garden State amber. Scientists at the American Museum of Natural History have discovered tiny mushrooms preserved in the Cretaceous amber from New Jersey. The mushrooms are virtually identical to the modern genus *Marasmius,* the little parasol mushroom. At present there are some eighty specimens known. Mushrooms are the fruiting bodies of fungi that grow underground as a mass of small threads known as mycelia; they are not really plants because they do not possess chlorophyll and hence cannot produce their own food through the process of photosynthesis. Instead, they depend on decomposing plant matter for nutrients. They are thus an important part of forest food webs, because they help recycle dead organic matter back into the soil, where it may be taken up and used again by other growing plants.

Could scientists possibly clone a dinosaur from the blood preserved in a Cretaceous blood-sucking insect in amber, as depicted in *Jurassic Park?* Aside from the fantastic complexity of such an undertaking, the prospect is unlikely for several reasons having to do with the nature of the specimens themselves. Although blood-sucking insects did indeed exist that might have fed on dinosaur blood (Cretaceous mosquitoes have been found in amber from New Jersey), our chances of getting one with a fresh meal in its gut seem vanishingly small. Also, even though amber is a very good preservative, there are chemical reactions occurring between the amber and the inclusions inside it that would lead to the breakdown of complex molecules like deoxyribonucleic acid (or DNA), the essential ingredient of cloning. Finally, there is often a tiny system of cracks penetrating the amber, as well as internal gas bubbles, both of which could supply the oxygen that would fuel decay reactions leading to at least partial destruction of the contained DNA. Some insect specimens in amber are seen to be in the initial stages of decay.

Nevertheless, the DNA from organisms trapped in amber may prove useful in establishing evolutionary relationships. In a technique known as DNA matching, molecular biologists have compared the DNA sequences of organisms to determine how closely or distantly they may be related. This process has been successfully applied to insects in amber from younger deposits[10] and may prove useful in determining the family lines of these earlier insects from the Raritan and Magothy beds of New Jersey. Despite this promising line of research, our DNA-cloning technology has a very long way to go before it would be possible to reproduce a living dinosaur in the laboratory.

Unfortunately, the evidence for dinosaurs themselves is limited in these New Jersey strata. There is but one bone known from this stratigraphic interval in New Jersey, and the only other proof is in the form of footprints.

Years ago, a fragment of a dinosaur bone was found in a clay pit at Roebling along the Delaware River in Burlington County. Precise stratigraphic data is lacking for this specimen, but the area in which it was found is underlain by Potomac, Raritan, and Magothy beds, making the fossil about 100 million years old and thus the oldest known dinosaur bone from New Jersey. It was collected by Dr. C. C. Abbott, a famous nineteenth-century archaeologist, who donated the specimen to Princeton University. Dr. Donald Baird has identified the fragment as the second metatarsal (footbone) of the right foot of a carnivorous dinosaur similar to *Albertosaurus*, a smaller tyrannosaur that lived later in the Cretaceous.[11]

The other hints of dinosaur life in middle Cretaceous New Jersey came in the form of a series of discoveries of trackways in 1929 and 1930. The first set of tracks consisting of four footprints was uncovered by workmen in the Hampton Cutter Clay Works pit at Woodbridge in Middlesex County in January 1929. Despite the best efforts of the New Jersey State Museum, the tracks were destroyed, but not before they were photographed and sketched in the field by Meredith Johnson of the New Jersey Geological Survey. His pictures and notes record the imprints of a large, three-toed, bipedal dinosaur. A year later, in January of 1930, a second trackway was uncovered at the same pit. Paleontologists from Rutgers University were called in, and this time a single footprint was successfully removed. This track was preserved and is today on display in the Rutgers Geological Museum in New Brunswick (see figure 6.3). This was one of four footprints in the trackway; again the other three impressions were apparently destroyed.

A third set of tracks was discovered in March of 1930. This time a host of geologists and paleontologists descended on the Hampton Cutter pit, and Katherine Graywacz of the New Jersey State Museum renewed her efforts to obtain the trackways. Plaster caps were applied to the carefully excavated footprints to protect them during removal; four tracks were apparently removed, while a fifth was destroyed in the process. Dr. Barnum Brown

FIGURE 6.3 Dinosaur footprint from the Raritan Formation. Found at the Hampton Cutter clay pit near Woodbridge, Middlesex County, N.J. Note three-toed track with deeply imprinted heel. This is the sole surviving specimen of the carnosaur trackway unearthed in 1929–1930. Courtesy of Rutgers Geology Museum, New Brunswick, N.J.

of the American Museum of Natural History, the famous dinosaur hunter and celebrated discoverer of *Tyrannosaurus rex,* came to Rutgers University to inspect the find. He identified the footprints as those of a large carnivorous dinosaur and noted that they were the only known Cretaceous dinosaur footprints from east of the Mississippi River, a distinction that still stands today.

The last set of footprints removed from the pit was to have become an exhibit at the New Jersey State Museum, with copies going to the American Museum, the Smithsonian Institution, and the Yale Peabody Museum. This never happened, and the fate of the four removed footprints is a mystery. The only surviving footprint from all these finds is the single track excavated in January 1930 and now on display at Rutgers. From all the remaining evidence, however, Dr. Donald Baird has reconstructed the trackways and made some further inferences about them.[12] He believes that all of the tracks were found on the same surface and that they were part of the same trackway made by one animal. They were discovered in the Woodbridge Clay Member of the Raritan Formation, making them about 90 million years old. The preservation of the footprints is due to their original impression into a bed of firm clay, which was quickly buried by a layer of sand. The footprints were around four feet apart, with the midline of the trackway passing through the base of the inner toe prints; this indicates a dinosaur walking upright with its legs tucked in directly beneath the body. There was no trace of a tail drag mark between the footprints, so the dinosaur must have walked with its tail held up off the ground. The footprints themselves were twenty inches long from middle toe tip to the base of the heel. The toes end in pointed claw marks; there is even evidence, on the remaining Rutgers footprint, of a backward-pointing "spur" or hallux, the impression of the vestigial first (or "big") toe. Baird compared these footprints to dinosaur tracks found in Germany, Australia, British Columbia, Texas, Israel, and Spain. He concluded that the New Jersey tracks resemble the widespread Early Cretaceous carnosaur tracks of "megalosaurian" type rather than any dinosaur footprints of later age.

Although dinosaur fossils have proven elusive in the Raritan and Magothy deposits, these beds should be considered promising targets for future prospecting. They are sediments that were laid down on land and in freshwater environments, in places where dinosaurs would have lived. It is a curious feature of the fossil record, however, that where tracks are found, bones are often not present. Footprints may be the best evidence we can hope for in terms of potential new dinosaur discoveries in the sands and clays laid down in the middle of the Cretaceous Period in New Jersey.

CHAPTER SEVEN

CRETACEOUS SEA LIFE

Much of New Jersey once lay beneath the ocean, and the evidence is in the marine fossils found in the state's coastal plain. In the Raritan and Magothy Formations in Middlesex County, there are layers that contain marine fossils of shellfish and sharks' teeth. In the overlying Cretaceous sedimentary formations in the Atlantic Coastal Plain area, the deposits record a repeated sequence of sea-level changes as the height of the ocean rose and fell. This oscillation of the sea produced intervals of continental flooding that alternated with periods of draining, like a great tidal cycle that washed over the land and then ebbed away. In the American Midwest this resulted in the inundation of the plains region by an inland sea that stretched from the Gulf of Mexico northward to the Arctic Ocean (figure 7.1). Geologists call this ancient sea the Western Interior Seaway, and its marine deposits are thousands of feet thick. Contained within the marine beds are the remains of the animals that swam in and flew over the stretch of ocean waters that were at times a thousand miles wide.

These inundations left their mark in the materials that make up the inner part of the coastal plain in New Jersey today: greensand marls, clays, and sands (figure 7.2). The soft sediments register the record of sea-level fluctuations as well as preserve the fossils of sea life that thrived in the Cretaceous waters.[1] The greensand marl (or glauconite as it is known scientifically) represents the flooding stage, while the clay and sand were deposited during the long ebbing stages. The pattern of alternating deeper water and shallower water deposits was recognized as early as 1907 by Stuart Weller, largely on the basis of the shellfish fossils found in these beds.[2]

The most abundant fossils in marine deposits are often the ones that cannot be seen, at least not with the naked eye. The ultimate source of all sustenance in the oceans are the microscopic plants that float in the surface

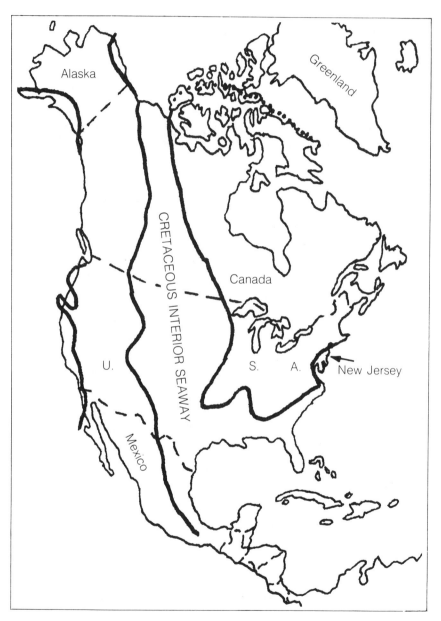

FIGURE 7.1 Map of North America in the Late Cretaceous Period. The middle of the continent was flooded by the Interior Seaway (after W. B. Gallagher, *Dinosaurs: Creatures of Time*, New Jersey State Museum Bulletin No. 14 (1990, Trenton, N.J.)

FIGURE 7.2 Map of Cretaceous formations, Inner Coastal Plain of New Jersey (stippled area).

currents of the seas. These tiny phytoplankton are the foundation of the marine food web, because they produce food from the raw materials of nutrients and water using the energy of sunlight. Although they are minute, their numbers are so great that they represent the largest store of organic mass in the ocean. Feeding on them are the minuscule floating animals known as zooplankton. The zooplankton form the crucial link in the marine food chain between phytoplankton and the larger creatures that eat them; big fish eat little fish, and so on.

In the Cretaceous seas, the most abundant plant plankton were the microscopic algae called coccolithophorids, which formed shells out of calcium carbonate (lime) dissolved in seawater. Although their name is much bigger than they are, these organisms bloomed in such tremendous numbers in the warm Mesozoic waters that after death the platelets from their lilliputian shells formed thick layers of chalk on the sea floor. In fact, it is the widespread deposits of chalk which gave the Cretaceous Period its name; Cretaceous is derived from the Latin word for chalk. This fine-grained limestone can be seen dramatically exposed in the White Cliffs of Dover in England, or in the badlands of western Kansas, where it outcrops as the Niobrara Formation.

The foraminifera were the major zooplankton group of the Late Mesozoic ocean. They too made their little shells out of lime. They can be thought of as amoebas that secrete a shell. These shells are very distinctive for each different species, and the forams (as they are called by their friends) changed rapidly over time so that they are very useful as index or guide fossils for correlation. Some forms floated in the upper currents of the ocean, feeding on phytoplankton and on each other, while other kinds lived on the sea floor.

There was no dearth of animals to eat the planktonic food resources of the Cretaceous oceans. One group of marine animals that eats plankton simply sits in one spot and filters the water for the food particles that it contains; such organisms are called suspension feeders. Clams, mussels, and oysters are suspension feeders, and the Cretaceous seas had abundant populations of bivalves, some of them very large as clams go. Big clams called inoceramids grew flat shells up to a yard long. This snowshoe design prevented them from sinking in the soft murky bottom of the chalk seas. Smaller inoceramids have been found in New Jersey in the Mer-

FIGURE 7.3 Cretaceous strata of the New Jersey Coastal Plain (chart).

GROUP	FORMATION	FOSSILS
Monmouth	Tinton	clams, snails, crustaceans, ammonites
	Red Bank	clams, snails
	Navesink	clams, snails, ammonites, vertebrates
	Mount Laurel	clams, snails, crustaceans, vertebrates
Matawan	Wenonah	clams, snails
	Marshalltown	oysters, reptiles, sharks
	Englishtown	clams, burrows
	Woodbury	clams, snails, ammonites, dinosaurs
	Merchantville	clams, snails, ammonites, worm tubes
	Magothy	plants, clams, snails
	Raritan	plants, insects in amber, dinosaur tracks
Potomac		plants, dinosaur bones (in Maryland)

chantville Formation at places where that clay deposit was mined for bricks, such as at the old Graham Brickyard pits in Maple Shade.

Some clams grew to resemble large sponges. Rudists, as these oyster relatives are known, grew one large barrel-shaped valve, covered at the top by a flat valve. This acted as a lid that could be opened for feeding. These bivalves cemented themselves together, forming reefs in the warm shallow seas. They thrived in the tropical latitudes of the Cretaceous. Today, the buried rudist reefs are important subsurface petroleum reservoirs in the oil fields around the Gulf of Mexico.

Large thick-shelled oysters like *Exogyra* and *Pycnodonte* were the most common bivalves on the Cretaceous sea floor in New Jersey and elsewhere in the Atlantic coastal region. These oysters had a convex valve that rested in the bottom mud and a flat valve that was positioned at the sea-floor surface, camouflaged by a thin layer of green mud. They needed thick shells and disguise to protect them, for just as in modern waters there were many animals around that enjoyed a meal of raw oysters on the half-shell. Other smaller oysters called *Agerostrea* clustered together, cementing shell

FiGURE 7.4 Collecting Cretaceous oysters in the Navesink Formation at Poricy Brook, Monmouth County, N.J.

to shell; these types had curlicued shells. The Cretaceous oysters formed dense banks that were preserved in the greensand marl beds of New Jersey; they are especially numerous in the Mount Laurel and Navesink Formations of Monmouth County, where the fossiliferous layers are exposed by stream erosion (see figure 7.4).

Many different varieties of snails crawled around on the Cretaceous sea floor (figure 7.5). Some forms ate the algae growing on the bottom, an old tried and true way to make a living for gastropods, as biologists address them. But in the Cretaceous, new kinds of snails appeared, such as whelks, which were carnivorous. They attacked clams and oysters by drilling into their shells or by forcing them open with their thick muscular foot.

Today, the molluscan group known as the cephalopods ("head-foot") has been reduced to a few shell-less forms such as the octopus and the squid. In the Mesozoic oceans there was a much wider variety of shelled types resembling the modern chambered nautilus of the Indo-Pacific region.

FIGURE 7.5 Representative Cretaceous marine invertebrate fossils. Top left, *Choristothyris plicata* (a brachiopod), front, side, and back views; top right, sponge (*Cliona cretacica*) borings in fossil oyster shell; middle left, *Gyrodes abyssinus* (a drilling carnivorous snail), top view; middle right, *Anchura pennata* (an herbivorous grazing snail), side view; bottom left, *Turritella encrinoides* (a high-spired herbivorous snail), side view; bottom right, *Pyropsis trochiformis* (a low-spired carnivorous snail), top view and side view. All specimens actual size.

The animal grew a coiled or straight shell, which was divided into chambers by internal walled partitions. The last and largest chamber was open at the end and contained an octopuslike animal with eyes, tentacles, and a beaked mouth. This organism had a jet-propulsion system and moved through the ocean by expelling a stream of pressurized water through a muscular nozzle called a hyponome. The cephalopod could control its buoyancy and hence its position in the water column by regulating the amount of gas in the empty chambers of the shell. The most abundant cephalopods in the Mesozoic marine realm were the ammonites, named after the ram's horn symbol for the ancient Egyptian deity Ammon-Ra.

Ammonites underwent bursts of evolutionary radiation during the Cretaceous that produced numerous varieties in relatively short spans of geologic time. They were active swimmers whose populations ranged over wide areas of the ocean. These features of their history have made them excellent guide (or index) fossils for establishing age equivalence (or correlation) over long distances. For example, we believe that some of the Western Interior Seaway deposits of South Dakota were laid down at about the same time as the New Jersey marl beds, because their contained ammonite fossils are so similar.

The rapid evolutionary changes in this group of mollusks reveal themselves in the structures of the ammonite shell. Perhaps the most distinctive structures are the suture lines formed where the edge of the internal chamber wall meets the external shell surface. In the modern chambered nautilus this line is a simple smooth curve, reflecting the simplicity of the curved chamber wall. In ammonites, however, the chamber wall became complexly crenelated, creating complicated suture lines that appear as intricate designs preserved on the fossil shell's surface. Each ammonite species has its own unique suture pattern. Some ammonites also developed ornamentation on the exterior of the shell, such as knobs, spines, or ribbing; the purpose of this sculpture was probably at least in part defensive. Additionally, ammonites coiled or uncoiled their shells in many different ways, some of them somewhat bizarre in appearance.

What this means practically is that each group of beds in the New Jersey Cretaceous marl deposits has its own characteristic ammonite assemblage. The Merchantville Formation, for example, has an ammonite fauna of large, tightly coiled flat-spired forms with complex pointy sutures (*Placenticeras*)

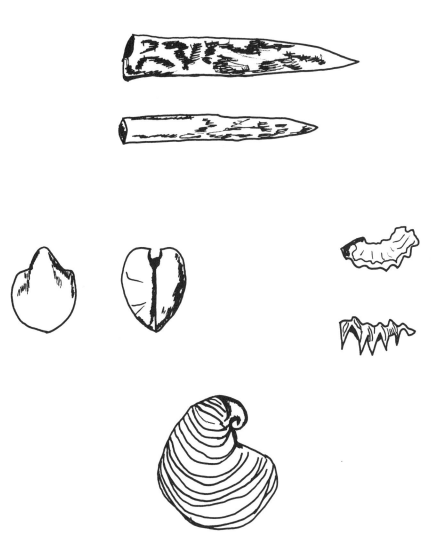

FɪɢᴜʀE 7.6 Representative Cretaceous marine invertebrate fossils. Top, *Belemnitella americana* (squid pens), side view; middle left, *Granocardium tenuistriatum* (a cockle shell), front view and side view; middle right, *Agerostrea falcata* (a small scalloped-shell oyster), top view and side view; bottom, *Exogyra cancellata* (a larger thick-shelled oyster), front view of right valve. All specimens actual size.

and smaller more openly coiled ornamented types (*Scaphites*). In the younger overlying beds such as the Mount Laurel and Navesink Formations, one can find an ammonite fauna dominated by the more unusually coiled heteromorph ammonites. One such heteromorph, *Anaklinoceras* from the Mount Laurel Formation of Delaware, has a shell that was tightly coiled in youth, grew straight for a while, then openly coiled back on its original shell. What purpose this pattern of development served is not entirely clear; perhaps the animal spent more time on the sea floor as it matured.

For one brief interval of Cretaceous time in New Jersey the most abundant cephalopods were the prehistoric squids known as belemnites. Their cylindrical amber-colored shells, called squid pens by some collectors (see figure 7.6), are found in tremendous numbers in the Mount Laurel and Navesink Formations. Unlike modern squids, the belemnites secreted a hard shell within which the animal encased itself in a hollow chamber. Judging from the copious squid pens found in these deposits, the belemnites swarmed in plentiful schools above the ancient oyster banks. They disappeared as suddenly as they appeared, perhaps as a result of changes in the ocean's temperature.

Crustaceans were also abundant in the Cretaceous waters. Ghost shrimp, lobsters, and crabs are found as fossils in the Atlantic Coastal Plain marl beds. Ghost shrimp claws in particular are very common, and the burrows of these crustaceans are preserved in nearshore sand deposits as rusty-colored branching structures covered with nodes that give the vertical tubes a distinctively bumpy surface texture.

With such an abundance of shellfish around, it is not surprising that there were a variety of larger vertebrate predators prowling the Cretaceous waters. Many of these predators were shellfish-eaters, as we can tell from their broad blunt teeth. New Jersey's Cretaceous deposits contain the fossil flat teeth of skates and rays as well as the blunt teeth of primitive spiny shell-eating sharks called hybodonts (see figure 7.7). By the later part of the Cretaceous Period, more modern breeds of sharks had appeared. These types, known as lamnoid sharks, had sharp pointy teeth and a vertebral column that was strengthened by the addition of calcium phosphate to the cartilage. The lamnoids were very successful, eventually giving rise to most of the present-day families of sharks. Some of the Cretaceous formations have great numbers of shark teeth in them, reflecting the fact that sharks

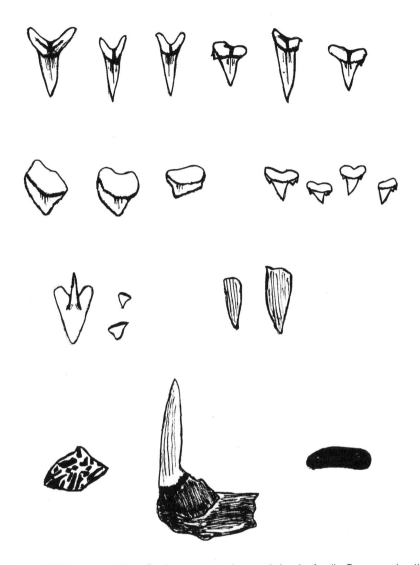

FIGURE 7.7 Representative Cretaceous marine vertebrate fossils. Top row: teeth from the goblin shark *Scapanorhynchus texanus*, lingual views, showing variation in the shape of the teeth dependent on position in the jaw; second row left, teeth of the extinct crow shark *Squalicorax pristodontus* (first two teeth) and *Squalicorax kaupi* (third tooth from left), lingual views; second row right, teeth of *Cretolamna appendiculata*, a prehistoric relative of the mako shark, lingual view; third row left, a large cephalic hook from a hybodont shark with two small teeth of *Hybodus*, an extinct spiny shark; third row right, two teeth from the bulldog tarpon *Xiphactinus*; bottom row left, a scale from the large garpike *Atractosteus*; bottom row middle, a piece of jaw with the enlarged palatine fang of the bony fish *Enchodus*; and bottom row right, a "bean" tooth of the shell-crushing fish *Anomaeodus phaseolus*. All figures actual size.

continually shed and replace their teeth, and attesting to the durability of hard enamel teeth as fossils. In some streams in Monmouth County, currents have concentrated shark and ray teeth in the stream bed and point bar gravels. Collectors wade into streams like Big Brook and Poricy Brook and dig up the stream gravels, placing them in homemade sieving boxes with bottoms made of window screen; washing the gravel removes the mud and sand, leaving behind a concentrated residue rich in sharks' teeth, oyster shells, and squid pens. The sharks' teeth can range in size from barely visible to impressive fangs several inches long. But sharks may not have been the fiercest predators in the Cretaceous ocean; that distinction probably belonged to other animals.

True bony fish swam in the greensand seas. Mostly these were primitive types like sturgeons. There were a host of now-extinct forms, like the "coral-nibblers," pycnodont fish with thin but deep bodies and blunt bean-shaped teeth that mark them as shell-crushers. The most common bony-fish teeth in the Cretaceous coastal plain deposits belong to a distant relative of the salmon called *Enchodus*. *Enchodus* was a small- to medium-sized fish with disproportionately large fangs at the front of the lower jaw (figure 7.7). Two very large teeth protruded from the front of the mouth, while much smaller teeth followed them in the rest of the jaw. Since its big fang is often found associated with fossil reptile skeletons, it is probable that *Enchodus* was a marine scavenger, taking advantage of large deadfall carcasses on the sea floor.

Even larger teeth and a few jaw fragments suggest the presence of the great bulldog tarpon *Xiphactinus*. This fourteen-foot-long fish had a blunt muzzle filled with sharp teeth, giving it a very pugnacious appearance. Other teeth of similar form but with serrations on one edge belong to extinct sailfish. A variety of smaller fish were also present.

But the real rulers of the Cretaceous sea were the marine reptiles. All of the large marine predators were reptiles whose ancestors had returned to the water, probably to take advantage of the rich food resources of the Mesozoic oceans. Among these were some animals that are familiar to us today. Turtles belong to an ancient lineage that dates back at least as far as the first dinosaurs. Originally the earliest turtles were land animals in the Triassic Period; by the Late Jurassic turtles had taken up a marine existence. In the late Cretaceous, sea turtles underwent a burst of diversification, perhaps because of the widespread warm ocean waters. Side-necked turtles, such as *Bothremys,* were abundant in the estuaries and shallow nearshore waters. These unusual turtles, known scientifically as pleurodires, are now freshwater animals restricted to the southern continents, but their thick fossil shell pieces are common in some New Jersey Cretaceous marine deposits. Pleurodires can withdraw their necks into their shells in a sideways motion, different from the vertical neck flexure of most familiar turtles; this is why pleurodires are known as side-necked turtles.

Other forms more closely allied to the familiar present-day turtles were beginning to display the architecture that led to the modern sea-turtle shell. Sea turtles do not need as much heavy bony armor weighing them down, and so today's marine turtles have a shell in which the amount of bone has been greatly reduced to increase the buoyancy of the animal. This trend began in the Cretaceous, and it allowed the appearance of some very ample turtles. *Archelon* was a large marine turtle whose skeleton has been excavated from the Western Interior Seaway deposits in South Dakota; it grew to a length of twelve feet, with a shell that was almost eleven feet across. But most of this shell consisted of bony struts instead of solid bony plates. *Archelon* also had developed paddles for effective swimming. While *Archelon* has not been found in New Jersey yet, there are intriguing bits and pieces of large turtle fossils from the greensand beds that suggest that big turtles were present in the Cretaceous of the east.

Another familiar group of Cretaceous marine reptiles were the croco-diles. Today the saltwater crocodile of Australia and southeast Asia swims across open ocean stretches, and the American crocodile of the Carib-bean region will also venture out to sea. In the Mesozoic, there was a diverse variety of crocodilians plying the ocean waters. As with the turtles, the croc-odiles started out as a terrestrial group in the Middle Triassic, appearing at about the same time as the earliest dinosaurs. By the Early Jurassic, fully marine forms known as teleosaurs are found in the Liassic rocks of Europe. In the Cretaceous, a true giant appeared: *Deinosuchus,* the "terror croco-dile." This fearsome reptile grew to an estimated thirty feet in length, with a huge gnarled snout that contained a mouthful of stout teeth. Such an ani-mal, as Drs. Donald Baird and John Horner have observed, was capable of taking a dinosaur for its dinner.[3] *Deinosuchus* inhabited tidal backwaters and estuaries where the sea waters mingled with mangrove swamps; duck-billed dinosaurs would have lived in this environment, taking advantage of the rich plant food resources that grew along the coastal marshes. Lurking in the dark tidewaters, *Deinosuchus* was ready to ambush its unsuspecting prey by lunging suddenly from under the surface, to grab its victim and drag it backward, thrashing and rolling as it retreated into the deeper water. Some fossil concentrations found in brackish water deposits of Cretaceous age may represent the accumulated remains of crocodile victims, especially where dinosaur teeth have been found stripped of their enamel, a characteristic of denticles that have been processed by the strong stomach acids of crocodiles.

Other Cretaceous marine reptiles are more or less familiar. The adherents of the supposed Loch Ness Monster will claim that it is one of the long-necked sea reptiles known as plesiosaurs. Although it is hard to imagine a Mesozoic marine reptile persisting in a Scottish loch that was until about ten thousand years ago covered by a glacier, at one time the plesiosaurs were very widespread and abundant in the oceans of the world. Plesiosaurs probably descended from the small Triassic aquatic reptiles called nothosaurs, which had long necks and streamlined bodies with shortened legs that were not yet flippers. By Early Jurassic time, the true plesiosaurs had evolved. Nicely preserved specimens from the bituminous shales of Europe display the small head, long neck, heavy body, short tail, and four flippers that typify the early plesiosaur body plan. By the Cretaceous, plesiosaurs had diverged into two distinct groups. One family, the elasmosaurs, grew very long necks with small heads at the end. This was the type of plesiosaur whose skeleton Professor E. D. Cope had mistakenly mounted with the head on the end of the tail. The other major group, the pliosaurs, was bull-necked with a large stout head. The bodies of both types were essentially the same, with a turtle-shaped torso, four flippers, and a medium-sized tail. The shoulder and hip bones of the plesiosaur were very sturdy and platelike, providing a broad flat surface of support on the animal's underside; this structure may have been useful if the marine reptile had to haul out onto land in order to lay its eggs, the way sea turtles come up onto the beach today to dig nests and lay eggs. The stout girdle bones also served to anchor the strong swimming muscles used to power the flippers; plesiosaurs paddled through the water in a kind of subaqueous flight similar to the way penguins swim today.[4] In the late Cretaceous marine deposits of New

Jersey, fragmentary remains and partial skeletons provide evidence of a short-necked plesiosaur called *Cimoliasaurus*.[5] Plesiosaur fossils have been found in Middlesex, Monmouth, Burlington, and Gloucester Counties in New Jersey. *Cimoliasaurus* and its pliosaur cousins were probably pursuit predators, chasing down their prey in the Cretaceous waters; the long-necked elasmosaurs were ambush predators, capturing fish by striking out with their long snakelike necks.

But as the Cretaceous Period went on, plesiosaurs became less common. They were receiving some effective competition from another group of reptiles more recently returned to the sea: the mosasaurs, giant marine lizards that first appeared around the middle of the Cretaceous. From similarities in the skull anatomy, mosasaurs are thought to be related to the modern monitor lizards, including the Komodo Dragon of Indonesia, at ten feet in length the largest living lizard. From their first appearance some 90 million years ago, mosasaurs diversified and spread rapidly to become the dominant marine predators of the Late Cretaceous oceans. They had a very open skull and long flexible jaws filled with sharp conical teeth. In the lower jaw, a joint in the middle allowed the animal to pop open its maw in order to swallow larger prey (see figures 7.8 and 7.9). In the back of the upper jaw was an additional set of teeth, the pterygoid teeth, which ensured that struggling prey would not be able to escape. The rear hinge between the back of the skull and the end of the lower jaw also seems to have been somewhat flexible.[6] Add to this armament a powerful body propelled by a long muscular tail and steered by four flippers, and you have a formidable carnivore capable of killing just about anything in the Cretaceous sea.

Mosasaurs came in a variety of types. While maintaining basically the same body plan, variation in size and in the anatomy of the head were the chief differences between genera. The older genera include *Clidastes,* a small fish-eater known from Matawan Group deposits in Maple Shade in Burlington County, Swedesboro in Gloucester County, and the Chesapeake and Delaware Canal in Delaware. Another mosasaur from this time interval is *Globidens,* a shell-crushing type with spherical teeth also known from the spoil piles along the Chesapeake and Delaware Canal near Summit Bridge in Delaware. The younger strata of Cretaceous age contain a greater variety of larger forms; the Monmouth Group deposits have yielded specimens

FIGURE 7.8 Mosasaur skull. This is the skull of *Mosasaurus maximus,* the largest and last of the mosasaurs. Skull length = 3.5 feet. From the Navesink Formation, Mantua Township, Gloucester County, N.J. New Jersey State Museum.

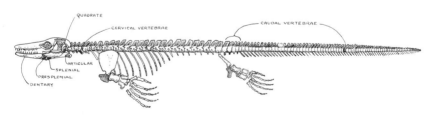

FIGURE 7.9 Skeleton of the mosasaur *Prognathodon rapax.* Based on a specimen from the Navesink Formation, Mantua Township, Gloucester County, N.J. The intramandibular joint in the lower jaw is between the dentary and the splenial bones; in the back of the skull, there was also movement between the quadrate and articular bones, allowing even greater flexibility in opening the jaws. After R. G. Chafffee, A New Jersey mosasaur of the subfamily Platecarpinae, Academy of Natural Sciences of Philadelphia, *Notulae Naturae* 137 (1939): 1–5.

FIGURE 7.10 Life in the Cretaceous ocean. In the foreground a mosasaur chases an ammonite; in the background are schools of bony fish and belemnites. Courtesy of S. Mulholland.

of the mosasaurs *Halisaurus, Liodon, Plioplatecarpus, Prognathodon,* and *Mosasaurus.*[7] Of these, *Halisaurus* is the smallest and shows the most resemblance to its monitor lizard ancestors. The largest of the mosasaurs was *Mosasaurus maximus* at some forty feet in length; two skulls of this animal have been found in the Navesink Formation of Gloucester County, New Jersey. Larger mosasaurs like *Mosasaurus* and *Tylosaurus* were capable of taking an assortment of prey items including ammonites, sharks, bony fish, birds, and even other smaller mosasaurs. Ammonite shells have been found with puncture marks that match the tooth pattern of mosasaurs.[8] In other evidence for mosasaur diet, mosasaur skeletons with the abdominal area intact have revealed stomach contents containing the bones of the animals mentioned above.[9] The mosasaurs were the top predators in the late Cretaceous marine food chain, eating just about whatever they wanted (figure 7.10).

In the air over the ancient seaways, the reptiles also ruled supreme. The largest flying animals were the pterosaurs, dinosaur cousins who had developed wings and taken to the air. The earliest forms are Triassic in age, and by the late Cretaceous some of them had become very large. *Quetzalcoatlus* from Big Bend National Park in western Texas is a late Cretaceous flying reptile whose wingspan is estimated at nearly forty feet across. A close relative of *Quetzalcoatlus* lived in New Jersey around the same time. A single delicate neck bone has been found in the Navesink Formation at Atlantic Highlands in Monmouth County. This indicates the presence of a large pterosaur in this area. A pterosaur neck bone and a wing bone have also been found in slightly older deposits along the Chesapeake and Delaware Canal in Delaware.[10] Pterosaur bones are very rare because they are hollow fragile bones, built for flight, that would be easily destroyed before burial. Since most pterosaur bones are found in marine deposits like the Navesink Formation, they are thought to have led a fish-eating way of life, swooping out of the open air over the sea to take schooling fish near the surface, much like a modern pelican. They may in turn have fallen victim to mosasaurs or other predators lying in wait for them near the ocean's surface.

CHAPTER EIGHT

THE LAST DINOSAURS

In the Late Cretaceous, the last part of the Age of Dinosaurs, the earth was a very different planet. An observer peering down from outer space would have seen a world unlike our own, one in which the outlines and positions of the continents and oceans did not much resemble the map of today's globe. A great seaway cut through the continents around the equator, an ocean that modern scientists call the Tethys Sea. Warm equatorial waters poured through what is now Central America and the Middle East, giving rise to blooms of plankton whose remains would someday provide the raw material for great subterranean reservoirs of oil. India was adrift in the ocean, an island continent whose collision with Asia was still millions of years in the future. North America, as we have seen, was flooded and cut into two by the Western Interior Seaway; a similar inland sea had invaded Europe. Everywhere the ocean encroached on the lowlands, in a series of sea-level rises thought to have been caused by the outpouring of molten lava onto the sea floor. The great volume of the hot rock displaced water onto the land. As the submarine eruptions ceased and the volcanic rock cooled and contracted, water gradually drained back into the ocean basins, until the next series of eruptions started the cycle all over again.

As sea level rose, Southern New Jersey was inundated, and the green-sand marl deposits formed on the shallow sea floor; when sea level fell, clays and sands were deposited as the ocean waters retreated. During this ebbing phase of the sea-level cycle, coastal swamps and marshes bordered the shore and migrated across the coastal plain as sea level changed. Estuaries formed where rivers met the sea, life teeming around the bays and barrier islands that made up the transition from land to marine environment. In this landscape New Jersey's last dinosaurs roamed, taking advantage of the abundant food resources along the coast.

The climate of the entire planet seems to have been generally warmer than it is today. This included New Jersey, whose sea waters were warm year round, perhaps owing to the ameliorating influence of the Tethyan currents. Fringing the shoreline were subtropical forests and swamps filled with a bounty of plant food.

While mosasaurs and plesiosaurs swam in the ocean, dinosaurs inhabited the adjacent coastal lowlands. Mosasaurs and plesiosaurs were not dinosaurs, however. All dinosaurs were land animals: the specific anatomical features that unite dinosaurs as a group are the hips and legs designed for land locomotion. Yet, in New Jersey, dinosaur bones are found for the most part in the same marine deposits that yield mosasaur and plesiosaur remains. This would seem to be a paradox; if dinosaurs were all terrestrial animals, why are their fossils found in marine beds?

To understand the answers to this question, we must turn to the gruesome discipline of taphonomy. Taphonomy is the branch of paleontology that is concerned with how living organisms become preserved as fossils. Taphonomists study how a skeleton can become disarticulated and buried. They also look very closely at the fossils to determine how they have been modified, for example by scavengers. The study of taphonomy also involves looking carefully at the bones in their sedimentary context, to see what clues they can provide us about how they came to be buried and preserved.

The most famous specimens of dinosaurs are whole skeletons. There are some well-known areas in western North America, such as the Red Deer River Valley in Alberta, where conditions were excellent for preserving entire skeletons of dinosaurs. In New Jersey, the best skeleton that has yet been found is the original specimen of *Hadrosaurus foulkii* (figure 8.1). With some forty-nine bones and teeth, this specimen was the most complete skeleton of a dinosaur known when it was described and named by Joseph Leidy in 1858. It would have been even more complete had not curiosity seekers removed some of the bones from the original discovery some twenty years earlier. As it was, Leidy and Foulke managed to recover forty bones or pieces of bone and nine teeth, plus some smaller fragments, which were obtained by sieving the clay around the bones. Of the skeleton, the most numerous bones obtained were the spinal bones, primarily the vertebrae of the neck (three), the back (seven), and the tail (eighteen). These showed

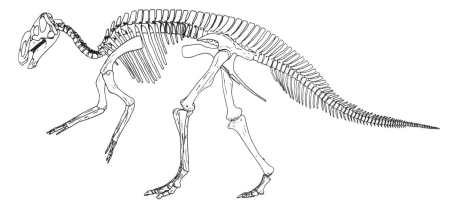

FIGURE 8.1 *Hadrosaurus foulkii* skeleton. Courtesy of Richard Rush Studio, based on reconstruction by D. Baird and J. Horner (from W. B. Gallagher, *Dinosaurs: Creatures of Time,* New Jersey State Museum No. 14 (1990), Trenton, N.J.

variability of the vertebral form along the spinal column, which is not unusual. Also recovered were the arm bones, the leg bones, half of the hip, a few foot bones, and a toe bone, all from the left side of the animal. The teeth were blunt but coarsely serrated, indicating a plant-eating animal. Leidy compared them to the teeth of *Iguanodon* from England, and to a tooth that he had named *Trachodon,* obtained from F. V. Hayden and collected in the Judith River badlands of Montana. The femur of the animal was massive, but the humerus (upper arm bone) was only half as long (see figure 4.1), so Leidy deduced that this dinosaur could stand on its hind limbs and go about in a two-legged fashion. The structure of the hip and leg bones also suggested this posture.

Hadrosaurus was found in a layer containing numerous well-preserved fossils of seashells, composed of original shell material so delicate that there was no question that they were buried where they had lived, without being transported. So *Hadrosaurus* had to come from somewhere else. How had it gotten to its final resting place? Today, when animals die along a river or stream, their carcasses can float downstream, buoyed up by gases of decomposition. Larger animals' cadavers will float quite far, sometimes out to sea. Here they will be subject to scavenging and sink to the bottom as the gases escape. They may be further scavenged on the bottom; in the modern marine realm deadfalls of large whales become a major resource for the animals in the ocean depths. Leidy's careful work produced some evidence in this direction; closely associated with the dinosaur skeleton the excavators found the teeth of a shark (*Scapanorhyncus*) and a tooth from a bony fish (*Enchodus*).[1]

Hadrosaurus gave its name to the group of duck-billed dinosaurs that are collectively called hadrosaurs. Hadrosaur bones are the most common kind of dinosaur bones found in marine deposits.[2] This has led some authorities to speculate that at least some hadrosaurs lived near the sea, in coastal environments where they inhabited the banks of streams and rivers, adapting a semiamphibious way of life. Some remarkable discoveries of mummified hadrosaurs from out west seem to lend credence to this idea. These specimens preserve intact the impression of the skin. They reveal that the hands of hadrosaurs were webbed. This is usually an adaptation for life in or around the water. Other scientists have speculated that the deep long tails of hadrosaurs may have been used as a powerful swimming oar. Yet recent discoveries in Montana have shown that some hadrosaurs nested in upland sites, away from the coast. It may be that hadrosaurs were socially complex animals that migrated to nesting areas during the breeding season, then returned to the coastal lowlands to take advantage of the lush plant growth. There is some evidence that they may have congregated in herds, and herding animals are often migratory.[3] Hadrosaurs were a diverse group of animals, and some species may have preferred upland areas, while others may have spent most of their time in coastal lowland conditions, where the ability to swim may have been advantageous.

Sometimes only a few bones are found together. Such was the case with an impressive discovery made along the shores of Raritan Bay, in Monmouth County. In 1869, Dr. Samuel Lockwood obtained some bones of truly massive proportions from the shore at Union Beach. He wrote to O. C. Marsh about his find, but before Marsh could act, E. D. Cope showed up on Lockwood's doorstep and insisted on seeing the bones. Lockwood tried to put him off, but Cope persisted and was granted access to the specimens. He quickly took measurements and sketched the fossils. In what may have been one of the opening salvos in the bone wars, Cope promptly published his description in the *Proceedings of the Academy of Natural Sciences of Philadelphia.* Marsh eventually purchased the bones and they reside today

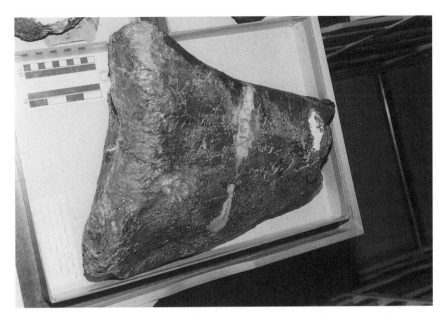

FIGURE 8.2 Ankle end of the tibia (shin bone) of "*Ornithotarsus immanis*" from Union Beach, Monmouth County, N.J. This specimen is from a large hadrosaur. Yale Peabody Museum.

in the collection of the Yale Peabody Museum, even though it was Cope who gave them the name *Ornithotarsus immanis*. Professor Cope determined that they were the lower part of the lower leg bones (or, in technical terms, the distal portions of the tibia and fibula) and the fused bones of the ankle (the astragalus and calcaneum). The name he gave them means "bird ankle," reflecting again the similarity of structure in the skeletons of dinosaurs and birds. But what is notable about these fossils is their truly impressive proportions. The long diameter of the tibia is over a foot across (figure 8.2), larger than the same bone in *Hadrosaurus foulkii* or many of the hadrosaur skeletons subsequently found out west. The measurements of the bones suggest that a truly enormous hadrosaur was present in Late Cretaceous New Jersey. Just how large, we cannot say for certain, since all that is known of this animal are a few associated bones.

Sometimes all we can find are single isolated bones, and some of these discoveries offer corroboration for the existence of a huge hadrosaur in New Jersey. One such specimen is the partial femur of a big duck-billed dinosaur

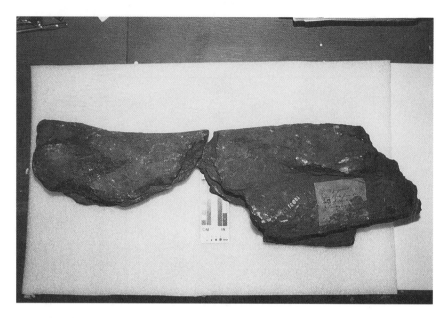

FIGURE 8.3 Giant hadrosaur femur from Swedesboro, Gloucester County, N.J. Academy of Natural Sciences of Philadelphia.

found in Swedesboro, Gloucester County, New Jersey (figure 8.3). The fossil thigh bone is in two sections, which fit together to form about half the length of the femur; if the entire bone were present it would be five feet long.[4] The specimen was presented to the Academy of Natural Sciences by a Mr. David Ogden; it was formally described by Joseph Leidy in his comprehensive 1865 summary of fossil reptiles from the United States. Its precise origin is unknown, although it is likely that it was found in one of the small marl pits around Swedesboro, possibly in the outcrop belt of the Matawan Group deposits. Today, the two pieces reside in the Academy's paleontology collections under the catalogue numbers 10001 and 10002.

These single bone occurrences are the primary record of Late Cretaceous dinosaurs in New Jersey. They are tantalizing clues about the nature of dinosaurs that hint at what might be found yet in the Garden State. For instance, in 1896 Lewis Woolman of the New Jersey Geological Survey reported the discovery of a dinosaur bone during excavation of a railroad cut for Pennsylvania Railroad right-of-way in Merchantville, Camden County,

New Jersey.[5] He took the bone to E. D. Cope, who identified it as the middle metatarsal or foot bone of the right rear foot of *Ornithotarsus immanis,* the giant hadrosaur Cope had named years before. This foot bone is fifteen inches long; the same bone of *Hadrosaurus foulkii* is only ten inches long. Scaling up from this one bone (admittedly a speculative exercise) suggests a hadrosaur some thirty-three feet long. The stratigraphy of the site where the bone was found indicates that the specimen came from the top of the Merchantville Formation, making it a little older than the original *Hadrosaurus foulkii* specimen, which was found only a few miles away but slightly higher up in the sedimentary stack. Today, excavations for highways, business parks, and housing developments may reveal fossils, including dinosaur remains, if the digging takes place in the right sedimentary deposits.

But perhaps the most common occurrence of dinosaur remains is in concentrations of vertebrate fossils in thin lenses or layers where disassociated pieces of many different kinds of animals may be found jumbled together. These thin layers may contain sharks' teeth, bony-fish scales and bones, turtle shell fragments, crocodile remains, and dinosaur teeth and bones. The mixture emphasizes the differences in the environmental preferences of the animals whose fossil remains accumulated in the same deposit. Sharks and dinosaurs did not live together in the same environment at the same time, so how did these anomalous collections of bits and pieces come to rest together? Taphonomists look for some sort of concentrating mechanism that could have brought together the teeth and disarticulated bones of disparate animals and buried them beside each other. Many of the bones display evidence of rolling, abrasion, and breakage, suggesting transport some distance. Thin layers of concentrations of this sort are often found where the sediment changes rapidly in composition both vertically and horizontally. What these concentrations seem to represent is deposition in estuaries or nearshore shallow waters during times of changing sea level. Perhaps concentrating and mixing energy was supplied by storms, which can move and rapidly bury organic remains in the proper environment. Thin concentrations of this kind are typical of the Monmouth Brooks fossil sites in Monmouth County.

These kinds of fossil concentrations are the most productive site for modern fossil hunters, now that the old marl-mining industry of nineteenth-

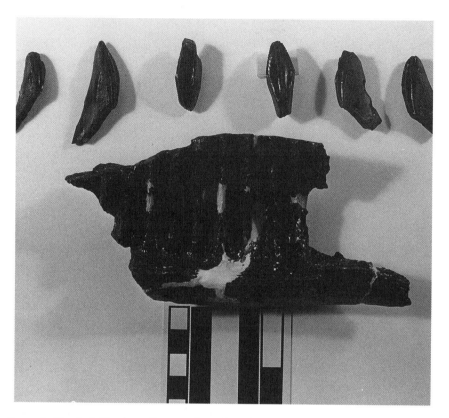

FIGURE 8.4 Teeth (at top) and jaw section of *Hadrosaurus foulkii.* Woodbury Formation, Haddonfield, Camden County, N.J. Academy of Natural Sciences of Philadelphia.

century New Jersey has largely disappeared. The brooks that cut through these beds will frequently erode the fossils out of the thin layers and then concentrate them as a placer deposit in gravel bars and deltas. Dinosaur teeth, bone fragments, and entire isolated small bones can be sieved out of the point-bar deposits along with sharks' teeth and the fossils of other animals. Perhaps the most common dinosaur remains found this way are the teeth of hadrosaurs (see figure 8.4). Toe and finger bones are also relatively stout and may survive short transport down a stream; other bones will tend to break, if they are not broken already as the result of their preburial journeys. Some other dinosaur fossils found this way are the teeth of carnivorous

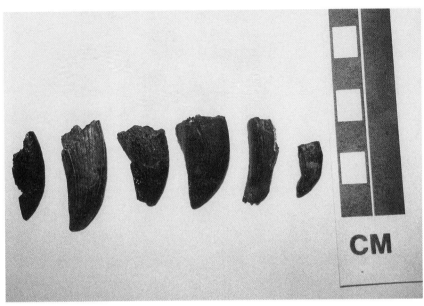

FIGURE 8.5 Teeth of meat-eating dinosaur, possibly *Dryptosaurus*. From Marshalltown Formation, Upper Freehold Township, Monmouth County, N.J. New Jersey State Museum.

dinosaurs (figure 8.5), fragmentary long bones (probably limb bones) of hadrosaurs, duck-billed dinosaur vertebrae (often broken), and the occasional piece of hadrosaur skull bone.

Two of the most unusual dinosaur fossils were found at the Monmouth Brooks sites. They represent our only two records thus far of armored dinosaurs or ankylosaurs in New Jersey. Ankylosaurs were dinosaurian tanks that were covered in bony armor; heavy-bodied and squat, they walked about on all four legs. A dorsal osteoscute or piece of bony armor from the back of an armored dinosaur was found at Big Brook near Marlboro, Monmouth County, and donated by its collector, Ralph Johnson, to Princeton University. The bony armor is distinctive enough to identify its former bearer as a nodosaur, a kind of more primitive armored dinosaur without a bony club on the end of its tail. A second discovery was that of a nodosaur vertebra or backbone, found in the stream at Poricy Brook near Middletown. Unfortunately, the osteoscute fossil was stolen from the Princeton University

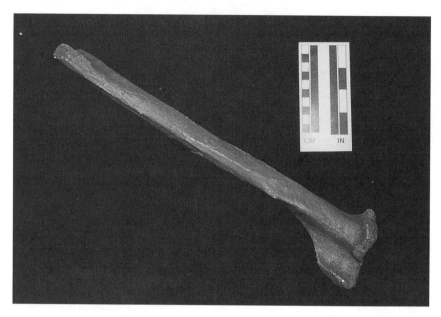

FIGURE 8.6 Tibia (shin bone) of *Ornithomimus (Coelosaurus) antiquus,* a hollow-boned small theropod from Burlington County, N.J. Cast of specimen in the collection of the Academy of Natural Sciences of Philadelphia.

collection by some unscrupulous individual, but plaster casts were made of it before it disappeared, so we still have a record of its form. The nodosaur vertebra resides in the collection of the Yale Peabody Museum.

Meat-eating dinosaurs were present in the area during the Late Cretaceous, and they appear to have come in two sizes. Smaller, delicate hollow bones found in both New Jersey and Delaware indicate the existence of a more lightly built carnivorous dinosaur. The first such specimen to come to the attention of science was a complete but hollow lower leg bone (tibia) from the greensand deposits of Burlington County (figure 8.6). This was described and named *Coelosaurus antiquus* by Dr. Leidy in 1865.[6] Leidy included in this description some bones that were brought to his attention by George Cook, the first state geologist of New Jersey and the man for whom Cook College of Rutgers University is named. Cook had wide-ranging interests in agriculture, soil science, and geology, and he also kept his eye out for unusual fossils as he pursued his official duties. Cook presented Leidy with a set of bones from a marl pit in Marlboro, Monmouth County, New Jersey. These included the bones of the lower leg and foot (the tibia plus two metatarsals and three toe bones; see figure 8.7). Leidy thought that it was a larger animal than the creature represented by the Burlington County tibia, and so had his doubts about whether or not the set of bones from Marlboro should be assigned to the same species.

Later, Leidy's former student Cope would give the Marlboro bones another name: *Laelaps macropus,* uniting the specimen with the carnivorous dinosaur Cope described and named *Laelaps aquilunguis* in 1866.[7] The species name refers to the observation made by Cope that, although the Marlboro tibia was smaller than that of his "eagle-clawed leaper," the bones of the foot were relatively larger than *L. aquilunguis.* Subsequently both the Burlington County tibia and the Marlboro bones have been referred to as *Ornithomimus antiquus,* because of the similarity of these remains to the bird-mimic dinosaurs known from more complete skeletons found in Cretaceous deposits of western North America. The bird-mimics were smallish (ten to twelve feet long) toothless theropod dinosaurs that looked very much like plucked ostriches with long tails, and arms instead of wings (figure 8.8). We now know that a wide variety of small carnivorous dinosaurs with hollow bones roamed the Cretaceous landscape, and it is difficult to determine to which group the foot and leg bones from New Jersey actually belong.

Isolated bones of small predatory dinosaurs continue to be found, and it is interesting to speculate about what kind or kinds of animals they represent. The leg bones when found are very hollow with only a thin layer of dense compact bone, a structural condition seen today in the bones of birds. Other distinctive bones are the phalanges (hand and toe bones; figure 8.9), which generally are short cylinders flat on one end and rounded into articulating condyles on the other end, with deepened pits on either side of the terminal condyles for insertion of the ligaments that worked and flexed the digits, including the claws at the end of the digits. One such claw has been found in the Marshalltown Formation sediments along the Chesapeake and Delaware Canal in Delaware. It is a short little stubby claw, unlike the great claw of *Dryptosaurus* (*Laelaps*). There are also small carnivorous dinosaur teeth that are sometimes found in the brook gravels, and these show that some very small predatory dinosaurs were present in Cretaceous

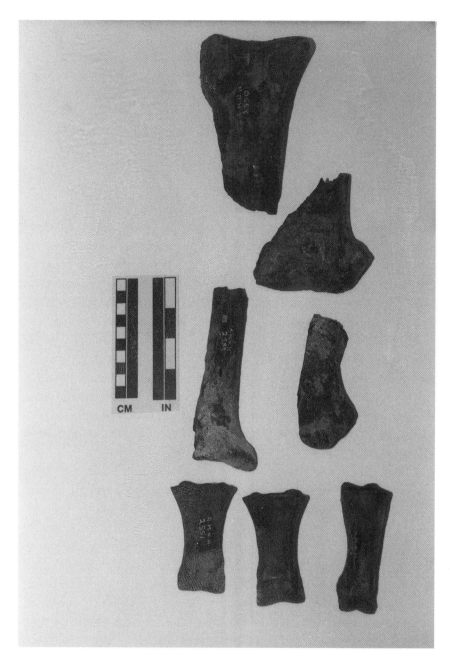

FIGURE 8.7 Leg, foot, and toe bones from a small theropod dinosaur *Coelosaurus antiquus* Leidy = *Laelaps macropus* (Cope) from Marlboro, Monmouth County, N.J. American Museum of Natural History.

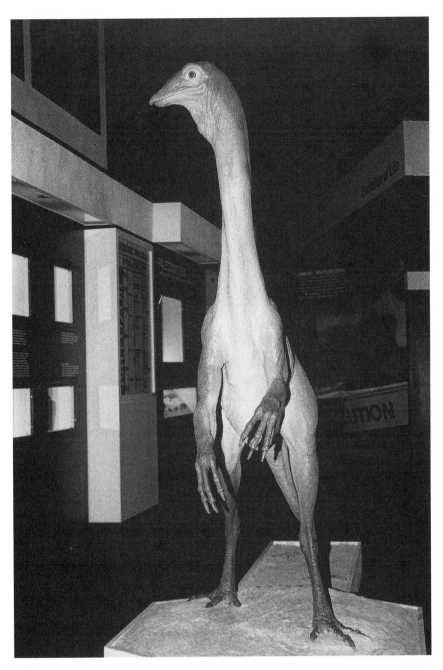

Figure 8.8 Model of *Ornithomimus,* the bird-mimic dinosaur. New Jersey State Museum.

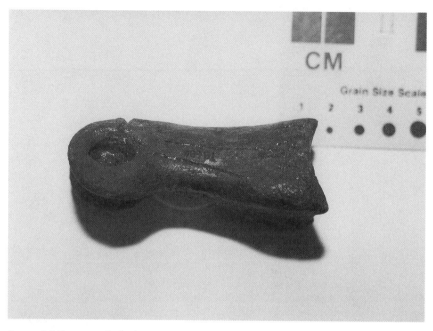

FIGURE 8.9 Theropod phalange (or toe bone) possibly from *Dryptosaurus*. Note pit on distal (left) end for insertion of ligaments that flexed toes and claws. From Marshalltown Formation, Upper Freehold Township, Monmouth County, N.J. Donated by Barbara Grandstaff. New Jersey State Museum.

New Jersey. Carnivorous dinosaur teeth are typically finely serrated along both edges (figure 8.10). Some teeth hint at the presence of raptors; NJSM 14664 is a cast of a tooth found in Mount Laurel, Burlington County, by Edward Lauginiger, and may possibly represent a kind of small predator like *Velociraptor.*

The picture we are left with suggests a population of small to medium-sized meat-eaters that were very lightly built, agile animals. It is not impossible that some of what we see of the fossil record of theropod dinosaurs represents growth stages of the same kind of animal; the somewhat imperfect sample may be a mix of the young and adult populations of several kinds of carnivorous dinosaurs. This problem may be solved by future discoveries, especially of more complete skeletons; unfortunately, the fragile nature of the hollow-boned remains makes them more susceptible to destruction before burial, especially if these remains have been subjected to

FIGURE 8.10 Close-up of *Dryptosaurus* tooth showing serrations along edges. Photograph courtesy of Robert Denton, Jr.

the kinds of energetic concentrating processes that seem to be responsible for the most prolific vertebrate fossil deposits in our area.

The best skeleton of a carnivorous dinosaur from New Jersey is again, like *Hadrosaurus foulkii,* one of the earliest. Leidy had reported on the teeth of a carnivorous dinosaur brought back to him from Montana in 1856, which he called *Deinodon.* Then in 1865 he described a larger serrated tooth found at Mullica Hill in Gloucester County and named it *Tomodon horrificus* (figure 8.11). Leidy thought it might belong to a plesiosaur, but later it was recognized as a carnivorous dinosaur tooth, in fact the largest that has yet been found in New Jersey.[8] But it would remain for Cope to demonstrate the existence of large predatory dinosaurs in New Jersey, and he did this in 1866 with his description of *Laelaps,*[9] the dinosaur whose name was later changed to *Dryptosaurus* by Marsh.

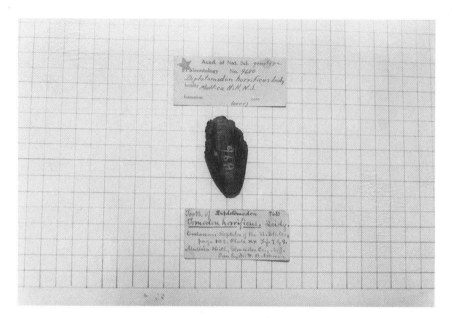

FIGURE 8.11 *Tomodon* (later called *Diplotomodon horrificus*), the first known carnivorous dinosaur tooth from New Jersey. From Mullica Hill, Gloucester County, N.J. Academy of Natural Sciences of Philadelphia.

FIGURE 8.12 Caudal (tail) vertebrae of *Dryptosaurus (Laelaps) aquilunguis* from Navesink Formation, Barnsboro, Gloucester County, N.J. Academy of Natural Sciences of Philadelphia.

FIGURE 8.13 Hand claw of *Dryptosaurus (Laelaps) aquilunguis,* with articulated phalanx (finger bone). From Navesink Formation, Barnsboro, Gloucester County, N.J. Academy of Natural Sciences of Philadelphia.

This partial skeleton consisted of sections of upper and lower jaws, several teeth, hourglass shaped vertebrae (mostly from the tail; figure 8.12), both upper arm bones (humeri), a complete set of leg bones, ankle and foot bones, and a number of phalanges, including the most remarkable feature of this creature, a large curved claw. The sharp, serrated, pointed teeth identify this dinosaur as a meat-eater, but its most formidable weapon must have been the great claw, which measures some eight inches in length (figure 8.13). Cope estimated a total body length of seventeen feet for the New Jersey meat-eater. The bones are very long and gracile, and some are quite hollow inside; Cope envisioned a very active predatory animal, and so he gave it the name *Laelaps aquilunguis,* meaning "eagle-clawed leaper." But while Cope believed that the large claw was located on the foot, today it is believed that this was a hand claw.[10]

Cope emphasized the bird-like aspects of *Laelaps,* and in some ways he anticipated the modern view of dinosaurs with his reconstructions of *Dryptosaurus* (*Laelaps*). Since Cope's day, numerous smaller, lightly built predatory dinosaurs have been discovered and described, and our ideas about dinosaurs have undergone some modifications. Some authorities have suggested that dinosaurs were warm-blooded animals like birds and mammals. They have cited as evidence the anatomy of dinosaur skeletons, which has prompted paleontologists from Richard Owen to Robert Bakker to imagine dinosaurs as very active animals different from modern cold-blooded reptiles in their behavior and metabolism.[11] Evidence for dinosaurian warm-bloodedness is equivocal; while some dinosaurs were undoubtedly built for speed, not all dinosaurs were capable of rapid movement,[12] and studies of dinosaur bone structure indicate a degree of metabolic activity somewhat intermediate between modern reptiles and birds.[13]

Cope's first dinosaur bone was found only a few feet away from the *Dryptosaurus* specimen in the pits at Barnsboro. This bone (see figure 4.4) came from the same stratigraphic level, the uppermost part of the "chocolate marl" at the West Jersey Marl Company excavations. It was clearly a portion of a hadrosaur thigh bone (femur) although substantially smaller than the same bone of *Hadrosaurus foulkii*. The knee joint was preserved, and the bone itself was about two feet long, but the end that fit into the hip socket was missing. Because it was found so close to *Dryptosaurus* (*Laelaps*), Cope was convinced he had found the same predator-prey relationship that was supposed for the English *Iguanodon* and *Megalosaurus*.

While there is abundant evidence for very large hadrosaurs in Cretaceous New Jersey, there also seems to have been a smaller species of duck-billed dinosaur in the area toward the end of the age of reptiles. In 1870, Marsh described several vertebrae of a smaller hadrosaur obtained from the same excavations at Barnsboro.[14] He named these bones *Hadrosaurus minor,* to distinguish them from the larger *Hadrosaurus foulkii*. In 1948, Edwin Colbert of the American Museum of Natural History reported on a hadrosaur found in a marl pit only a few miles away from the old Barnsboro mines.[15] This was a partial skeleton (see figure 9.1) of a duck-billed dinosaur that was about three-quarters of the size of *Hadrosaurus foulkii*; the vertebrae from this specimen compare closely in size to Marsh's vertebrae, and Colbert concluded that there was a smaller duck-billed dinosaur species in the younger Cretaceous deposits of New Jersey. In fact, *Dryptosaurus aqui-*

FIGURE 8.14 Hadrosaur jaws: baby jaw (top) and adult jaw (bottom). From Manalapan, Monmouth County, N.J. American Museum of Natural History.

Iunguis and *Hadrosaurus minor* are the last dinosaurs known from the Cretaceous beds of New Jersey. There are no more dinosaurs above the level at which these dinosaurs are found.

There is even some evidence for baby dinosaurs in New Jersey's Cretaceous deposits. From the same West Jersey Marl Company workings, Marsh received the tiny shoulder bone of a very young duck-billed dinosaur. This bone is still in the Yale Peabody Museum under the number YPM 7898. Other bones from New Jersey indicate the presence of juvenile dinosaurs. A tiny jaw fragment in the collections of the American Museum of Natural History is from Manalapan in Monmouth County, New Jersey. An identical bone of much larger size is a fragment of hadrosaur jaw from the same spot (see figure 8.14). Shed hadrosaur teeth found in the Monmouth Brooks fossils sites come in smaller and larger sizes, and some are quite diminutive. As we saw above, some of the smaller carnivorous dinosaur teeth from New Jersey may represent younger growth stages of the larger meat-eaters. Indeed,

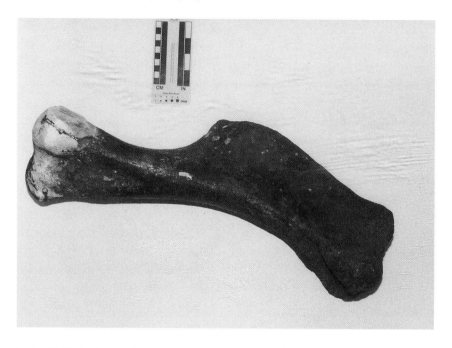

Figure 8.15 Hadrosaur humerus (upper arm bone) from Monmouth County, N.J. This is the specimen whose dimensions suggest that it may be from a crested lambeosaurine hadrosaur. Academy of Natural Sciences of Philadelphia.

Cope noted that the skeleton of *Dryptosaurus* (*Laelaps*) from Barnsboro was that of an immature individual because of the incomplete fusion of some of the bones.

There are hints at some other dinosaur types in the Garden State. For instance, ANSP 15550 is a large duck-billed dinosaur humerus, the bone of the upper arm, from Monmouth County. The bone is possessed of a large and prominent ridge running down the front half (figure 8.15); in true flat-headed hadrosaurs, this ridge is smaller than the diameter of the shaft of the bone. But in this particular specimen, the bony ridge (or deltopectoral crest) is as large as the diameter of the shaft, a feature more characteristic of the crested duck-bills known collectively as lambeosaurines. These hadrosaurs had a variety of bony protuberances sticking out of their heads, projections of the nasal bones that contained enlarged nasal passages. The function of these crests has been the subject of much speculation, and ideas put forth to explain them include uses as an air supply chamber while diving, enhanced smelling capability, and as a sort of bassoon that could produce low-pitched sounds. While a number of lambeosaurine species are known from the western states, the evidence for their existence in New Jersey is scanty. Aside from the specimen mentioned above, two other specimens (YPM 3216 and NJSM 11961) may pertain to the lambeosaurines.

Both of these specimens are of the forearms of hadrosaurs. YPM 3216 is a slender radius and ulna that was forwarded to Marsh from the West Jersey Marl Company pits at Barnsboro. These pits are still visible today in the Mantua Township Municipal Park near Pitman. The old excavations have filled in with water, forming several long ponds; the banks are thickly overgrown, and no outcrops of greensand are now accessible, although exposures of the marl are present along the banks of nearby Chestnut Run. The old pits were largely worked by hand and by hydraulic mining; narrow-gauge railroad tracks were laid down into the pits, and small steam engines were used to haul out the mined marl. There is a local story that one of the old pony engines got stuck in the mucky marl, and is still down there submerged under the stagnant pond water.

The other specimen, NJSM 11961, was collected in 1980, at the Inversand Company pit, from the upper part of the Navesink Formation. The specimen consists of the associated radius and ulna from the forearm of a duck-billed dinosaur (figure 8.16). The interesting thing about these bones is that the proximal ends are pathologically ossified; the elbow joint was a gnarled mass of bone. This dinosaur had an arthritic elbow. The pathological bone growth may have been due to an old injury that did not heal properly; perhaps the dinosaur had an accident and fell down, or perhaps it was wounded. At any rate, moving the arthritic arm was probably painful. In recent years, paleopathology has become a subject of interest to some students of dinosaurs, and many cases of broken ribs, tails, and other bones have been documented.

How many different kinds of dinosaurs were present in Late Cretaceous New Jersey? Many different names have been given to dinosaur remains from the Garden State, especially during the intense rivalry of the bone wars, when Cope and Marsh were trying to outdo each other in the naming of new dinosaurs. Some of these names were given to individual teeth or isolated bones, a practice that has fallen out of fashion with modern-day

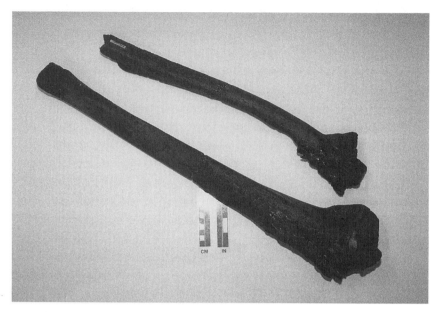

FIGURE 8.16 Hadrosaur radius and ulna (bones of the forearm) showing deformation of the elbow. The ends of the bones at the lower right of the picture are pathologically ossified, indicating that this dinosaur suffered arthritis in its elbow joint. From Navesink Formation, Mantua Township, Gloucester County, N.J. New Jersey State Museum.

paleontologists, who prefer to have more of the animal's skeleton before assigning a name. So for example Leidy named a theropod tooth from Mullica Hill *Tomodon* (later *Diplotomodon horrificus*); today we would probably refer that tooth to the carnivorous dinosaur *Dryptosaurus* (*Laelaps aquilunguis*). Other names founded on little material, such as *Ornithotarsus immanis* or *Laelaps macropus,* have been synonymized (or made equivalent) with dinosaur names based on more complete skeletons.[16]

But we can say with some degree of certainty that at least five different species of dinosaurs inhabited this area toward the end of the Cretaceous Period. There were two kinds of hadrosaurs, a larger form that lived earlier in the Cretaceous and a smaller species that lived later. There is rare but good evidence for the existence of an armored dinosaur, a nodosaur, in Cretaceous New Jersey. And finally there were at least two kinds of predatory dinosaurs, a smaller variety and a larger meat-eater with an outsized claw.

FIGURE 8.17 Adult *Hadrosaurus* with juveniles on a marsh in Cretaceous New Jersey. Painting by Larry Felder.

Hadrosaurus foulkii, Hadrosaurus minor, and *Dryptosaurus aquilunguis* are good names founded upon adequate skeletal material. Secure names for other New Jersey dinosaurs will have to await the discovery of more of their skeletons. But because of the chancy nature of the preservational process, complete skeletons are hard to find.

Above the level where the *Hadrosaurus minor* and *Dryptosaurus aquilunguis* skeletons were found, dinosaur fossils disappear altogether. In the younger overlying formations, no more dinosaur bones are found. A profound change came over the earth, and the nature of that change is still being debated.

CHAPTER NINE

THE GREAT EXTINCTION

Today, no large dinosaurs dominate the landscape. As nearly as scientists can pinpoint it, the last dinosaurs all died out about 65 million years ago. The causes for this disappearance have been the subject of much speculation, and in recent years an intense debate has arisen over the nature and rate of the extinction at the end of the Mesozoic Era. The change in the dominant life-forms at this point in geologic history was so striking to the pioneer paleontologists and stratigraphers that they decided to make it a major division between eras, similar to the changeover that occurred at the boundary between the Paleozoic Era and the Mesozoic Era, the mass extinction at the end of the Permian Period. Just as the Permo-Triassic mass extinction event had begun the age of reptiles, the Mesozoic Era, so the Cretaceous-Tertiary boundary (or K/T boundary, as it is abbreviated by geologists) initiated the Cenozoic Era and the age of mammals.

To be sure, many modern groups were already present at the end of the Mesozoic. As we have seen, flowering plants and many insects of modern aspect had appeared earlier in the Cretaceous. Birds were sharing the air with pterosaurs; primitive toothed birds like *Hesperornis* and *Ichthyornis* gave way before the end of the period to more modern toothless shore birds like plovers and curlews. Small, hollow, fragile bones of these shore birds are among the rarest of the vertebrate fossils found in the marl deposits of New Jersey.[1] Other small and delicate fossil remains found in the Monmouth Brooks sites indicate that in the underbrush various lizards, similar in form to present-day skinks, geckos, and iguanas, scuttled and basked. Along the stream banks and swamp margins, salamanders and frogs lived. Soft-shelled turtles essentially identical to today's trionychids were around, and the ancestors of modern sea turtles swam in the ocean waters. Crocodiles of various sorts were about, including small alligators. And hiding in the nooks and crannies of the dinosaurs' world were small mammals, little balls of fur that appear to have been nocturnal insect- and seed-eaters. Some of these were quite familiar in appearance; the teeth of *Alphadon*, a small Cretaceous opossum, have been found in the same deposits that produce dinosaur bones at one site in New Jersey.[2]

So a modern fauna of animals was present under the dinosaurs' feet. What then was the nature of the great changeover at the end of the Cretaceous? The disappearance of the dinosaurs is the most obvious difference between the Mesozoic fauna and the present day. Some very specific ideas ranging from the plausible to the ridiculous have been proposed to explain the extinction of the dinosaurs. One old idea is that the little mammals ate all the dinosaurs eggs. This proposal does not seem very likely, because at least some dinosaurs seem to have been careful parents that took care of their nests and their young.[3] Another reason that makes this idea seem less likely is that mammals and dinosaurs originated at virtually the same time in the Triassic, and mammals remained in the same ecological niches for the rest of the Mesozoic. Why didn't the mammals drive the dinosaurs into extinction earlier in their mutual Mesozoic history? And why did other egg-laying reptiles, such as crocodiles, turtles, and lizards, survive?

Another idea put forward to explain dinosaur extinction is that a worldwide plague spread through the dinosaur populations at the end of the age of dinosaurs. The problem with this one is that dinosaurs were a very diverse group of animals; most disease organisms only target a small group of animals, sometimes just one species. Would a disease that killed ceratopsians have also killed off hadrosaurs? Most disease organisms do not extirpate their entire host population, for a very good reason; it is bad evolutionary strategy, because killing all the host organisms then puts the disease organism out of business as well.

But there is something else wrong with these ideas and all the other theories proposed to account specifically for the extinction of the dinosaurs. The real problem with dinosaur-specific explanations of the K/T mass extinction is that this event involved the disappearance of a wide range of very different kinds of organisms, from the microscopic to the gigantic. The extinction event affected large populations of abundant animals in the oceans as well as on land. In fact, the extinction seems to have been even more severe in the seas. There was a major population crash of the coccoliths and forams, the microscopic oceanic plankton that is the base of the food pyramid in the marine realm. Large and abundant clam species were affected, and typically Cretaceous forms such as the inoceramids, the exogyrines, and the rudists dwindled and disappeared. The ammonites, so characteristic of Mesozoic marine faunas, all became extinct. Certain primitive shark types disappeared, like the shellfish-eating hybodonts. Primitive bony fish including *Enchodus* were also affected. Plesiosaurs and mosasaurs became extinct. In the air, the last of the pterosaurs vanished.

The mammals were also affected, but to a lesser degree. There were some extinctions among marsupials (pouched mammals) and multituberculates, which were primitive rodentlike mammals. But the main stocks of more advanced mammals were unaffected, perhaps because of their small size. An interesting feature of the K/T extinction is that all the larger animals on land were extirpated; one estimate is that all terrestrial animals weighing more than twenty-five kilograms (fifty-five pounds) died off.[4]

The extinctions were global in scope, affecting a wide variety of organisms in different ecosystems in an apparently short interval of geologic time. Just how long the extinctions took is at the crux of the debate over its causes. When we say "a short interval of geologic time," we are not being very specific. Geologic time is very long indeed, and a short interval could be interpreted as a few million years or hundreds of thousands of years, quite a long span of time on the human scale of things. On the human scale, a short interval might be days, weeks, months, or a few years. So the debate over the timing and causes of the K/T extinction has become an argument over the rate of extinction. Did all these different and diverse organisms disappear very rapidly, in a matter of weeks, months, or years, or was it a much more prolonged process that took hundreds of millennia, if not millions of years?

Linked to this question of gradual or sudden extinction is the question of the general source of the cause. Usually, theories of gradual extinction invoke some cause that is associated with the slow processes of geologic change. In these gradual scenarios, extinction is associated with some large-scale earthly process or processes that proceed at a geologic pace. Such a scenario would invoke an interlocked series of mechanisms such as an increased rate of mountain building, more volcanic activity, lowering of sea level, changes in ocean currents, and a deterioration of the climate at the end of the Mesozoic Era. All of these are gradual processes that would slowly change the earth's environment, making it incrementally inhospitable to those organisms adapted to the long summer of the Mesozoic. In this view, extinction is a process that occurs over millions of years and is driven entirely by terrestrial causes.

The opposite view has been termed catastrophism. In the catastrophist scenarios, the cause is usually single and sudden, and the extinction rate is very rapid. This point of view has come to be largely associated with those explanations for mass extinctions that rely upon an extraterrestrial cause. While catastrophism was espoused by such early luminaries of vertebrate paleontology as Baron Cuvier, the dominant mode of thinking in geology for the past two centuries has been gradualist, due to the influence of such early geologists as James Hutton and Charles Lyell. Only recently has geological science begun to consider seriously the role of catastrophes in earth's history, largely as a result of the debates over the nature and causes of mass extinction.

One explanation of the K/T mass extinction combines both earthly and extraterrestrial causes. Using the evidence from remnant magnetism in ancient rocks, geologists were able to trace the wanderings of the continents and figure out that they had all been connected in the supercontinent of Pangaea. But close study of the magnetic properties of rocks indicated that the polarity of the earth's magnetic field has also changed—not once, but many times in the prehistoric past. The magnetic field surrounds the entire planet and extends out into outer space, where it provides an invisible shield against incoming radiation from the sun and from the depths of space. The shifts in the earth's field are called geomagnetic reversals, during which the north and south poles switch positive and negative charges.

According to one explanation for mass extinction events, as the geomagnetic reversals occur, the earth's magnetic field weakens and the radiation from outer space pours in through the atmosphere, raising the radiation levels at the planet's surface to deadly levels. Radiation is not healthful to most organisms, and a brief lethal dose of radiation from outer space, according to this theory, is what killed off the dinosaurs and the other animals and plants that died out at the end of the Cretaceous.

The problems with this idea are twofold. First, the K/T boundary mass extinction occurred in the middle of an interval between reversals, far away in time from the reversal process itself. Secondly, water is an excellent shield against radiation and would protect at least some kinds of marine animals that we know became extinct.

Another hypothesis put forth was that a nearby supernova, the incredibly violent blast of a star blowing up, would bathe the earth in intense radiation. The blast could kill off organisms directly by radiation exposure, or indirectly by drastically altering the planet's weather. Such an explosion would have presumably left traces in the form of distinctive but very rare radioactive metals such as plutonium. Although some sediments from the K/T boundary layers have been tested for plutonium, no trace of the rare and deadly metal has been found.

But another interesting and rare element *has* been found at the K/T boundary, and this has led some scientists to propose a third extraterrestrial theory for the dinosaurs' extinction. While studying the pattern of geomagnetic reversals in the rocks of late Cretaceous and early Tertiary age exposed in a gorge near the town of Gubbio in Italy, geologist Walter Alvarez became aware of a thin but persistent layer of clay in what was otherwise a thick sequence of chalks. The chalks were marine deposits that contained the innumerable microscopic shells of planktonic plants and animals. When the chalks were studied closely layer by layer, it was clear that the Cretaceous forams and coccoliths disappeared right at the thin clay layer, while above the clay the microscopic plankton fossils were typical Tertiary forms.

Walter Alvarez thought this thin clay layer might yield a clue about how long the K/T extinction process had taken. He took some chalk and clay samples home to his father, the Nobel Prize winning physicist Luis Alvarez, who worked at the Lawrence Radiation Laboratories at the University of California at Berkeley. The Alvarezes decided to analyze the clay sample from Gubbio for the element iridium, a very rare metal belonging chemically to the platinum family of elements. Iridium is found in higher concentrations in meteorites, and a steady but small amount is supplied to the depleted earth's sediments by the dust of burned-up meteors, which settles through the atmosphere at a known and constant rate. The amount of iridium in the sample, the Alvarez team reasoned, should be a function of time. In most land and nearshore sediments this meteoritic dust is diluted by terrestrial materials, but in deeper water sediments the extraterrestrial dust can be detected by a high-tech process known as neutron activation analysis. This analytic technique is so sensitive that it is capable of measuring iridium at the level of parts per trillion, an extremely low level of dilution. The technique involves exposing the sample to a controlled concentration of free neutrons in a nuclear reactor for a specified period of time. Bombardment of the normally stable iridium makes it radioactive, and a given quantity of the unstable radioisotope will give off a specific and known quantity of energy in the form of gamma radiation.

The newly irradiated sample is placed in a lead-lined chamber containing a scintillometer, essentially a very sensitive Geiger counter that measures the energies of the rays being given off by the sample. The readings of the scintillometer are tallied up automatically by a computer, which produces a readout of the energy levels and a calculation of the concentrations of the element in the sample. In this way, very small amounts of rare elements may be measured with a reasonable degree of accuracy.

When the Alvarez team at Berkeley submitted the Gubbio sample to this procedure and got back the results, they were taken aback by the figures. Their tests indicated a level of iridium in the clay that could not possibly be explained as the normal slow small input of meteoritic dust. They had to come up with a different explanation.

In science, the process of generating an explanation for a set of observations is called formulating a hypothesis. In this case, the Alvarez team had to explain an iridium concentration in the Gubbio clay of 9.1 parts per billion (ppb), some thirty times the concentration of the rare metal found in the chalks both below and above the clay layer. In order to check whether the Gubbio samples were unique or representative of a larger phenomenon, they analyzed other localities with K/T boundary rocks for iridium. Samples from two well-known sites in Denmark and New Zealand showed substantially higher iridium levels at the K/T boundary (again, as at Gubbio, determined by the changeover in oceanic plankton fossils) than the levels in the surrounding rocks. The level of iridium was even higher in the Danish sample than the original Gubbio clay, peaking at 41.6 ppb.

These results suggested a world-encompassing event that occurred at a rapid rate. They first thought of the supernova theory as a possible explanation, but testing for plutonium revealed no trace of the deadly element, so they abandoned this idea. The hypothesis that the Alvarez team came up with to explain their results was the impact of a large body from outer space, probably an asteroid.[5]

From the level of the iridium concentrations in the K/T boundary sediments, they calculated that the object had a diameter of at least 10 kilometers (6 miles). The impact of such a large object traveling at a speed of 25,000 miles per hour would have caused a huge explosion and excavated a vast crater. The impact would have rapidly injected a very large volume of dust into Earth's atmosphere and this dust blocked out incoming sun light, severely restricting the ability of plants to make food through the process of photosynthesis. Once the plant food was gone, plant-eating animals would die out, taking with them the meat-eaters that depended on herbivores for food. This would have been as effective a killing agent on land as it was in the sea. The process was global and virtually instantaneous from a geological perspective, taking between several months to several years to occur. This particular catastrophist proposal became known as the Alvarez hypothesis.

The asteroid impact proposal immediately generated a great deal of interest and research across a broad spectrum of scientific disciplines. Researchers very quickly started supplementing the Alvarez team's original findings with reports of iridium spikes at K/T boundary sites all over the globe. Space scientists started looking at the data on the earth's craters, trying to find a likely suspect for the K/T killer. Theoretical models elaborated on the extremely destructive effects of a large asteroid impact on the planet. Other lines of evidence including widespread droplets of melted glass, quartz grains with shock features, types of silica only formed at very high pressures, and high levels of soot, all associated with the iridium-enriched layer, were offered as corroboration for the Alvarez impact hypothesis. The results of all this work were presented at several specially organized scientific meetings, and this interaction between scientists from diverse disciplines served to increase the excitement over the asteroid impact scenario.[6]

But the idea did not immediately win widespread acceptance among all members of the scientific community. Paleontologists especially were cautious about the impact idea, perhaps because of the inherent conservatism of the profession, instilled by the long tradition of Lyellian gradualist emphasis. Some skeptics pointed out that there were other ways to get lots of iridium on the earth's surface—for instance, from volcanoes such as those in the Hawaiian Islands that tap magma from deep in the earth's mantle where iridium is more abundant. Indeed there is much evidence for very large and widespread volcanic eruptions in the later part of the Cretaceous Period, and as we noted above intense volcanic activity is part of the gradualist scenario for the K/T boundary extinction.

Additionally, some paleontologists pointed out that the fossil record appeared to show that some groups were already declining and disappearing before the very end of the Cretaceous, a pattern more consistent with a gradualist explanation for the extinctions. For example in the northern High Plains of North America, at sites in the Hell Creek Formation, an iridium anomaly had been located in some sections in terrestrial floodplain deposits just at the K/T boundary. But in these same sections the closest dinosaur skeletons were located two meters (about six feet) below the iridium-enriched layer, implying some period of time between the last-known dinosaurs and the deposition of the iridium.

One major problem with the Alvarez hypothesis was the lack of a "smoking gun": where was the crater of the right age and of the right size for the hypothesized asteroid? Several candidates were put forward and found wanting, including the Manson structure buried under the glacial till of the Iowa prairie; at 35 miles across, it was not large enough to be the killer crater. The Alvarez team suggested that the impact had occurred in the ocean, and so was buried or otherwise obliterated long ago. But the paleo-detectives managed to deduce from the distribution of shocked quartz grains that the crater had to be somewhere in the southern part of North America, and that is where the search centered.

As is often the case with these things, someone had already identified the suspect, but the evidence was overlooked. Glen Penfield, a geophysicist working for the Mexican national oil company Pemex, had noticed an intriguing large-scale circular pattern on his sophisticated subsurface sensing equipment while looking for potential new oil fields in the Yucatan Peninsula. The gravitational and magnetic anomalies over the northern coast of the Yucatan indicated a buried circular structure over 120 miles across. Penfield decided to see if there were any well samples from old exploratory wells drilled in the area. All of the well samples had been destroyed in a warehouse fire except for a few that had been sent to New Orleans for further analysis. Alan Hildebrand, a Canadian geologist interested in impacts, managed to track them down. What he found was a set of well cores from a well drilled at Chicxulub in the Yucatan, near the center of Penfield's buried structure. The shallow rocks from the well were Tertiary limestones, not unusual for the Yucatan, but at depths hundreds of feet below the surface the well had penetrated an interval of glassy melted breccia, a kind of rock produced at high pressures and temperatures.[7] Melt breccia, as it has come to be known, was collected by Apollo astronauts from the surface of the moon. It has been interpreted as the product of meteorite impacts, although it may also form during very violent volcanic eruptions.

Here, the catastrophists thought, they had the smoking gun. The buried structure displayed all the characteristics of a large crater, including the central raised area seen in meteor craters on the moon. It was the right age and the right size for the K/T killer. Moreover, all around it in the Caribbean, in northeastern Mexico, and in the Gulf Coast of the United States, there was evidence in the rocks of a violent cataclysm at the very end of the Cretaceous. Chaotic rock beds associated with iridium spikes, shocked quartz, and glassy spherules could be seen in Haiti, at a place called Mimbral in Mexico, and along the Brazos River in Texas. These finds were interpreted by impact adherents as the remnants of the great tsunami, or impact-generated tidal wave, caused by the cratering event. As the great wave rolled across the shallow ocean floor, it tore up chunks of sea bottom and mixed them willy-nilly with the blown-out crater debris beginning to settle out of the atmosphere. As the water sloshed back from the land, it washed back plant material that was preserved in the stirred-up sediments. The great wave itself must be viewed as a major destructive consequence of the impact.

It would appear that the impact advocates had won the day. Prediction after prediction has been verified as a result of observations designed to test the Alvarez hypothesis. Still, some of the evidence seems ambiguous, and detractors of the impact explanation continue to insist that much of the data can be interpreted differently. The slow decline and disappearance of some fossil groups at the end of the Cretaceous is one piece of contrary evidence.

What does the fossil record in New Jersey tell us about this critical interval of geologic time? The sediments of the Monmouth Group in the Atlantic Coastal Plain contain fossils from the last part of the Cretaceous, while the

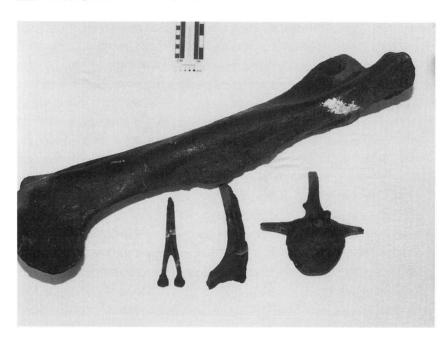

FIGURE 9.1 *Hadrosaurus minor* bones. These are from the level that produces the last and youngest dinosaur specimens in New Jersey. The large bone is the femur (thigh bone). Smaller bones below it are, from left to right, a chevron (tail rib); a portion of rib; and a vertebra. From upper Navesink Formation, Mantua Township, Gloucester County, N.J. Academy of Natural Sciences of Philadelphia.

overlying deposits of the Rancocas Group preserve the remains of early Tertiary organisms. The uppermost layers of the Navesink Formation in southern New Jersey have produced the last dinosaur remains, as we saw in the last chapter, including the partial skeletons of *Dryptosaurus aquilunguis* and *Hadrosaurus minor* (figure 9.1). In the bottom of the formation immediately above this, the Hornerstown Formation, the last specimens of mosasaurs, ammonites, and fishlike *Enchodus* and *Squalicorax* are found. In the upper part of the formation the dwarfed and scant fauna is composed entirely of small primitive forms that are only found in the Tertiary layers above. So on the basis of the fossils the K/T boundary would then be placed at the base of the Rancocas Group, in the basal part of the Hornerstown Formation (figure 9.2).

The last dinosaurs in New Jersey are about the same age as some of the

Figure 9.2 Upper Cretaceous to Lower Tertiary sequence of strata. This exposure of Late Cretaceous age and Early Tertiary age sediments brackets the K/T boundary in New Jersey. At the base of the pit, sediments of the Navesink Formation have yielded dinosaur bones; the overlying Paleocene strata of the Hornerstown and Vincentown Formations contain a Cenozoic fauna devoid of dinosaurs. Four-wheel-drive vehicle for scale. Gloucester County, N.J.

dinosaurs found in the lower part of the Hell Creek Formation in the northern High Plains region. We can tell this because an ammonite specimen recently discovered from below the level of the last New Jersey dinosaur discoveries is of the same species as ammonites found underneath the Hell Creek Formation in the Dakotas. At the time of the last dinosaurs, the sea was retreating and draining back off the continents. The change between the Navesink and Hornerstown Formations reflects this change in sea level. The last fossils of Cretaceous organisms are found in the greensand of the basal Hornerstown Formation, in very pure glauconite sands deposited in a rising sea. It has been argued that these fossils were actually removed from their entombing sediment and reworked into overlying Tertiary sediments, and some of them do appear worn or abraded; but many specimens are essentially complete skeletons, or articulated remains of such delicate nature that they could not have been moved without destruction.[8] At any rate it seems clear that we have a biological record of the last faunas of Cretaceous time.

As for iridium anomalies, the sediments of this interval have been tested, and no significant iridium spikes were found. However, the nature of the sediment itself makes it unlikely that such a spike would be preserved; the greensands have been thoroughly burrowed and eaten through by bottom dwellers that mixed the ancient mud so completely there is no trace left of the original layering.[9] Also, age estimates on the greensands do not help with the determination of the rate of the extinction process. The entire thickness of the Hornerstown (about eighteen feet) is estimated to have been deposited very slowly over an interval of about three million years. The Cretaceous fossil bed at the base is only about a foot thick, but it may have taken as much as several hundred thousand years for the sediment in that bed to be laid down. On the other hand, a catastrophic accumulation of remains could have occurred over some shorter period of time and then been buried.

What can we learn from the New Jersey fossils then? The fossil record can be a very good source of answers if the right questions are asked. If we look at the "big picture" of changes in the dominant animals in the Late Cretaceous and Early Tertiary, some interesting patterns begin to emerge. The dominant marine shellfish of Cretaceous marine waters were the large oysters such as *Exogyra* and *Pycnodonte.* The remains of these bivalves form very densely packed fossil shell beds in the greensands of Late Cretaceous age. Swimming in the water column over them were the ammonites, complex beautiful shellfish of many different varieties. Shell-crushing predators of many types including spiny sharks, skates, and primitive bony fish preyed on these abundant sources of seafood. Plesiosaurs ate the fish, while mosasaurs ate anything and everything, including the shellfish.

But the nature of the marine community in New Jersey changes dramatically in the fossil faunas of the upper layers of the Hornerstown Formation and into the overlying Vincentown Formation. The dominant marine organisms in terms of abundance are not the Cretaceous shellfish, but organisms like sponges, small solitary corals, and brachiopods. In fact, the brachiopod *Oleneothyris harlani* (see figure 10.3) is found in densely packed shellbeds at some sites in the upper Hornerstown and lower Vincentown formations. The Vincentown Formation is also known for its colonies of bryozoa, another animal not common in the Cretaceous deposits.

If you will think back to the beginning chapters of this book, you will remember that brachiopods were the dominant bivalved shellfish of the Paleozoic ocean, where they lived among communities of corals, bryozoa, and sponges. But after the Permo-Triassic extinction event that ended the Paleozoic, brachiopods were never really important in the Mesozoic seas. Somehow, after the K/T boundary, the evolutionary clock was reset, and the marine ecosystems were for a brief while in the Paleocene dominated by organisms of a more primitive Paleozoic aspect.

What was different about the Paleocene seas and the organisms that inhabited them? The least obvious difference is probably the most important one, and that is the absence of the abundant populations of diverse microscopic plankton typical of Cretaceous marine ecosystems. The great blooms of coccoliths were gone, and the forams reduced to a few small species. The ocean waters of the early Paleocene were food-poor as a result, compared to the cornucopia of plankton available to the Cretaceous shellfish. Moreover, many of the typical Cretaceous shellfish practiced a reproductive strategy that depended on the abundant planktonic food resources of the Mesozoic seas. The oysters and ammonites released their larvae into the waters to feed on the plankton and develop among them before metamorphosing into adult form. When the plankton suddenly died off (and their extinction pattern does seem abrupt), the shellfish dependent upon them for food both as adults and more critically as developing larvae were also affected.

Why then were brachiopods, sponges, and corals the most abundant invertebrates of Paleocene marine communities here in New Jersey? They are all minimalist organisms that have modest food demands. Their ancestors originated in the relatively impoverished Paleozoic seas, before the great plankton blooms of the Mesozoic. Living brachiopods, for example, have the lowest metabolic rate and lowest food requirements of any marine shelled invertebrate. Sponges are very simple organisms that are essentially colonies of loosely associated single cells. Moreover, none of these animals are necessarily dependent upon a larval stage in the plankton for their development to adulthood. They pursue different reproductive

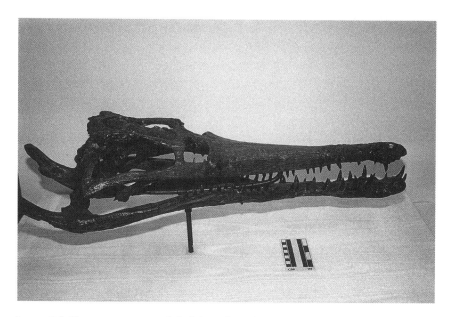

FIGURE 9.3 *Hyposaurus rogersii* skull from basal Hornerstown Formation, Mantua Township, Gloucester County, N.J. Crocodiles are examples of animals that survived the mass extinction event at the end of the Cretaceous Period. New Jersey State Museum.

strategies. Brachiopods, for instance, lay yolk-rich eggs that hatch out small but fully developed brachiopods ready to go to work as bottom-dwelling filter-feeders. Sponges can reproduce by budding, and corals undergo a phase of asexual reproduction as well.[10]

The marine vertebrate predators of the Paleocene were much diminished in size and variety from their Cretaceous predecessors. Shell-crushers were less common, and gone were the big marine reptiles like plesiosaurs and mosasaurs. Interestingly enough, the major benefactors of the extinction were the crocodiles, who were the major reptilian predators of the Paleocene (figure 9.3). The survival of the crocodiles after the K/T boundary extinction may be due to their cosmopolitan tastes; today crocodiles are nonspecialized predators who will eat anything from fish to human beings, and who inhabit a wide range of environments from freshwater inland waterways to the open sea. Generalists such as crocodiles seem to fare best during times of mass extinction.

What about the dinosaurs? The large land-dwelling dinosaurs were not at the peak of their diversity at the very end of the Cretaceous. Just some few millions of years before Hell Creek time, in the Judith River beds of Campanian age, the dinosaur fauna of the American West was flourishing, and dinosaurs were common elsewhere in the world. But toward the end of Hell Creek time, as the Mesozoic Era was drawing to an end, the variety of dinosaurs was reduced to a smaller number of large forms now well known to most schoolchildren, dinosaurs such as *Tyrannosaurus rex* and *Triceratops*. Some of these types, such as *Triceratops,* seem to have been abundant and may have roamed the latest Cretaceous plains of Montana and Alberta in large herds.

If we look at the animals that are disappearing today, the rare and endangered species, perhaps they can tell us something about the general nature of animals that go extinct. Conservation biologists have studied the dwindling populations of some of these endangered species, birds in particular, to learn how they might be saved from extinction. What population ecologists have determined is that smaller populations of larger species of birds have a greater risk of extinction.[11]

This has to do with several factors that are largely size-dependent. Large animals need a bigger range and more food resources to support each individual; for a given area of habitat their population will be relatively small. For instance, in any given acre of the polar region today, there might be thousands of little lemmings living, but only now and then will a single polar bear cross that acre. Large animals tend to reproduce more slowly than more prolific small animals. During times of environmental disturbance, the small populations of large animals will be more drastically affected, and it is more difficult for them to reestablish their previous population levels.

Dinosaurs, the ancestors of the birds, were probably subject to these same constraints. If we think of dinosaurs as giant bird relatives, then they were probably the most extinction-prone animals of all time. The probability of their extinction given any kind of environmental disruption was very high, since some of them were very large indeed. The latest Cretaceous dinosaurs were the largest and last of their families. In this, they followed a pattern that E. D. Cope had noticed in the fossil record of other groups. What has been known since as Cope's Law states that species within a lineage tend to grow larger and larger in size over time until the lineage goes extinct. By contrast, late Cretaceous mammals were all little animals, with larger populations of small individuals that had minimal resource requirements.

Of course, there are some modern paleontologists who argue that the dinosaurs never did become extinct. These scientists point out that if dinosaurs are the ancestors of birds, then the dinosaur line did not disappear, but in fact became even more successful, as the descendants of dinosaurs diversified and filled the air with all the different types of birds we see today.

CHAPTER TEN

AFTER THE DINOSAURS

The world of the early Tertiary Period was a place of opportunity. The disappearance of the dinosaurs left open many niches on land; in the ocean, similar possibilities existed. All of the large animals of the Mesozoic Era were gone, and there was plenty of room for other creatures to take their places. The little mammals who had hid in the shadows of the dinosaurs' world were the chief benefactors of the K/T mass extinction. They quickly moved into a number of habitats not open to them previously. But they did not immediately become the large predators of land and sea; that would take some time. In the meantime, other animals filled those roles.

In the Paleocene oceans, swept clean of the great marine reptiles, the way was clear for the sharks to become the dominant predators. They seem to have wasted little time. In the greensands of the upper part of the Hornerstown Formation, the earliest teeth of an ancestral great white shark, *Paleocarcharodon*, have been found. Closely allied to the great white shark, and found in the same layers, the teeth of *Otodus obliquus* represent another large carnivorous shark cruising the Paleocene waters of New Jersey.

Reptiles did not abandon the marine waters. Crocodiles, abundant as ever, seem to have been among the major carnivores of the oceans in the Paleocene. The most common crocodile in New Jersey's Paleocene was a narrow-snouted form called *Thoracosaurus*. Similar in appearance to the modern gharial crocodiles of the Ganges River in India, these ocean-going crocs were fish-eaters that grew to appreciable size.

But perhaps the largest of the Paleocene predatory reptiles was a snake, which paleontologists have named *Paleophis*. This serpent was a primitive constrictor related to the modern boa; its remains have been found in the Vincentown Formation and the Shark River Formation of New Jersey. On

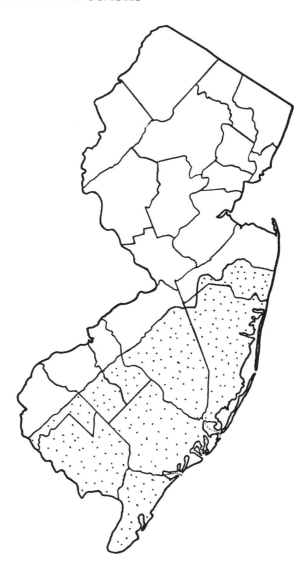

Figure 10.1 Map of Cenozoic Formations in New Jersey (stippled area).

the basis of the size of some of the ancient constrictor's vertebrae, E. D. Cope estimated its length at twenty feet.[1]

On the land, the most immediate inheritors of the dinosaurs' mantle were birds. Giant flightless birds with large sharp beaks and big talons on their feet were the largest terrestrial predators in the Paleocene and Eocene Epochs. A foot bone of one such bird, *Diatryma,* is known from the Eocene deposits of New Jersey.

The only evidence for land mammals in the Eocene of New Jersey was described by Joseph Leidy in 1867.[2] A single tooth was forwarded to Leidy from the marl mines around Shark River in Monmouth County, New Jersey. The precise stratigraphic origin of this tooth was somewhat equivocal; Leidy was uncertain about whether it came from the Eocene beds or from the Miocene deposits overlying them in that area. Leidy identified the tooth as that of a pachyderm, but subsequently it was recognized as belonging to a tillodont, a kind of extinct mammal whose fossils are found out west in Eocene rocks. Leidy gave to the animal that left behind this single tooth the name *Anchippodus riparius* and deposited the specimen in the collection of the Academy of Natural Sciences of Philadelphia. The tillodonts were a strange group of early Tertiary mammals that had clawed feet and long front teeth like rodents, but they were larger animals—in fact, one of the largest of the early mammal groups. They grew to about the size of a bear, although their diet was vegetation. The teeth, which grew continually, suggest that they were gnawers of tough plant material. Since tillodonts are only known from Early Tertiary rocks in the west, the New Jersey specimen is almost certainly Eocene in age.

Early whales were present in the oceans by late Eocene time, and one fossil vertebra in the collections of the New Jersey State Museum may be from an Eocene whale. It is again from the old marl pits along the Shark River, and while it is attributed to the Eocene Shark River Formation, there remains the possibility that it came from overlying Miocene deposits. The Eocene whales are better known from more complete specimens of large-toothed types like *Basilosaurus* from the Gulf Coast states. This was a long serpentine form, similar to a mosasaur in overall appearance, reaching lengths of eighty feet!

By the end of Paleocene time the diversity of marine invertebrates had reestablished itself after the decimation of the K/T boundary. In the Eocene, the marine shellfish fauna started to become more and more modern in appearance (see figure 10.3), and it was on the basis of the increasing percentages of marine invertebrate fossils that Charles Lyell subdivided the Tertiary Period up into smaller units of time named epochs (see figure 10.2).

Deposits of the next Tertiary time interval, the Oligocene Epoch, are

FIGURE 10.2 Cenozoic Rock Formations in New Jersey (Chart).

PERIOD	EPOCH	FORMATION	FOSSILS
Quaternary	Pleistocene	Cape May	clams, snails, land mammals
Tertiary	Miocene	Cohansey	rare seashell molds
		Kirkwood	mollusks, sharks, mammals,
	Eocene	Shark River	mollusks, shark teeth
		Manasquan	mollusks, microfossils
	Paleocene	Vincentown	bryozoa, coral, clams, sharks
		Hornerstown	brachiopods, coral, sponges

missing from outcrops of Tertiary beds in New Jersey, although there is some evidence that they are present in the subsurface, buried under younger strata toward the coast. Excellent exposures of Oligocene-age rocks are seen in the famous Big Badlands of South Dakota, where they form the heavily eroded and deeply incised badlands topography that presented a major barrier to settlers moving west in the nineteenth century. But because of this erosion, the fossiliferous beds of the Badlands have yielded a treasure trove of Oligocene land-mammal fossils which were first studied by Joseph Leidy in the 1840s. They reveal a fauna of weird mammalian types like *Brontotherium,* a large heavy-bodied herbivore with a flat, branching nasal horn. Although mammals ancestral to many of the more familiar modern forms, such as horses, were present in the Oligocene, the fauna was dominated by primitive groups, many of which are now extinct.

Miocene sediments are widely distributed along the eastern seaboard of the United States, probably because this was an interval of world-wide sea-level rise. Marine deposits are present in all of the Atlantic Coastal Plain states including New Jersey. In some places, these deposits have been mined for their phosphate content in vast pits dug out by giant drag-line shovels. At these excavations in North Carolina and Florida, the mining operations have uncovered a wealth of Miocene marine fossils. In other areas, such as the famous Calvert Cliffs of Maryland, erosion has created cliffs over a hundred feet tall, revealing layer after layer of Miocene shell beds. Bones and teeth erode out of the cliffs and wash up onto the beach, making

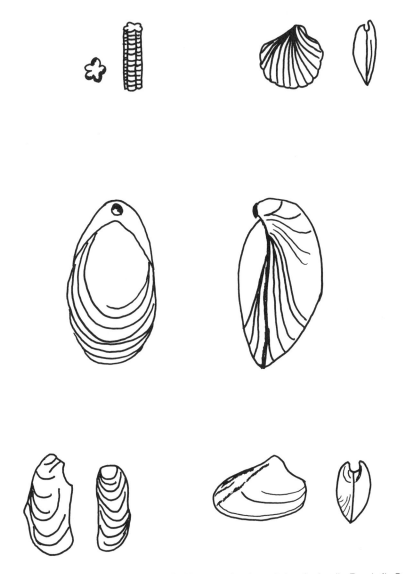

FIGURE 10.3 Representative Early Tertiary marine invertebrate fossils. Top left, *Pentacrinus bryani* (a crinoid from the Paleocene Vincentown Formation), top view and side view; top right, *Venericardia antiqua* (a clam from the Eocene Shark River Formation), front view and side view; middle, *Oleneothyris harlani* (a brachiopod from the Paleocene Vincentown Formation), front view and side view; bottom left, *Gryphaeostrea vomer* (a small oyster from the Paleocene Vincentown Formation), front view and back view; bottom right, *Etea delawarensis* (a clam from the Eocene Manasquan Formation), front and side views. All figures actual size.

beachcombing along the Chesapeake Bay in this area a very rewarding exercise. The waves constantly replenish the supply of cetacean bones and sharks' teeth along the strand line, and fossils out of the cliffs have given us a very good idea of the fecundity and diversity of Miocene marine life. Occasionally a rare land-mammal tooth is also found.

In New Jersey, most of what we know about Miocene fossils came out of the old marl-mining excavations, primarily from two areas; the marl pits around Shiloh and Jericho in Cumberland County, and the stream bank exposures and old marl mines along the Manasquan and Shark Rivers in Monmouth County. In the southern end of its outcrop belt, the upper part of the Kirkwood Formation was dug for marl until about World War I, when mining ceased in this area because of the discovery and exploitation of the richer phosphate fertilizers further south along the coastal plain. While mining was in progress, a rich fauna of marine fossils was obtained from the pits.[3] In addition to the abundant mollusc fossils, sharks' teeth and whale bones were also discovered in this deposit. There were a variety of sharks prowling the Miocene waters in New Jersey, including the prehistoric giant great white shark, *Carcharodon megalodon* (figure 10.4). This monster had teeth up to seven inches long and grew to fifty-two feet in length; it is estimated to have weighed up to fifty tons. Such a shark was bigger than any of the whales of its day and probably preyed on them. Modern great white sharks have been reliably reported at up to twenty feet in length.

The whale family diverged into two branches during the Oligocene; one became the odontocetes, which includes the sperm whales and the porpoises, while the other branch became the great filter-feeding baleen whales. Primitive shark-toothed porpoises like *Squalodon* were present early in the Miocene, although they were supplanted during the course of that epoch by more modern forms. The cetaceans evolved into a wide variety of small long-snouted toothed forms, many now extinct, whose vertebrae and tiny peglike teeth are their most common fossil remains. Small primitive baleen whales were also swimming in the Miocene seas of New Jersey.

Both invertebrate and vertebrate fossils are very similar to the fossils found in the Calvert Formation in the basal part of the Calvert Cliffs, which has been dated as late Early Miocene in age, approximately 15 to 20 million years old. The only land-mammal fossil known from the Shiloh marl is the tooth of a tapir, *Tapiravus*, a primitive semiaquatic hoofed animal related to

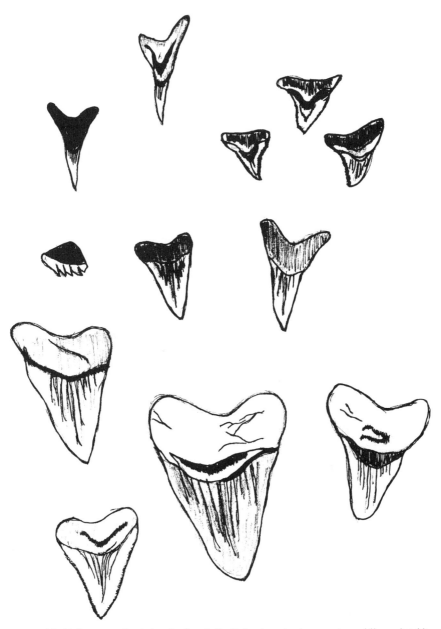

FIGURE 10.4 Miocene shark teeth. Top left, *Odontaspis elegans* (sand tiger shark); top right, *Hemipristis serra* (Indian Ocean shark); middle upper left, *Hexanchus* (cow shark or six-gilled shark); middle center, *Isurus hastalis* (mako shark); middle right, *Isurus desorii* (mako shark); middle lower left, *Isurus desorii* (mako shark); bottom row, *Carcharodon megalodon* (giant great white shark). All actual size.

the rhinoceros. Today, the tapirs are found only in the tropical rain forests of Central and South America and in Southeast Asia, where they live near rivers and streams. In the Miocene the tapirs were widely distributed across the northern continents.

In the central part of the Kirkwood outcrop belt, the formation consists of white, yellow, and salmon-colored fine sands and silts, and no marine fossils have been found here. However, excavations in this part of the Kirkwood have uncovered large petrified logs at places like Gibbsboro in Camden County. Pieces of petrified wood have also been found in the Kirkwood deposits at Blackwood and at Washington Township in Gloucester County. The wood is thought to have been from cypress trees, which today grow further south in swampy lowlands from Virginia to Florida.

Perhaps the most interesting of the Miocene fossils from New Jersey have come from the old marl workings and the stream bank outcrops in the Manasquan and Shark River areas of Monmouth County. From the old excavations, Leidy, Cope, and Marsh obtained the first New Jersey specimens of Miocene land mammals. One of the first specimens to appear was a peccary tooth from Shark River, which Leidy noted in 1867. Marsh then described two teeth found near Farmingdale and named them *Ammodon leidyanus.* These represented an interesting new animal whose real nature was not known until better specimens of the same kind were found in Oregon. Cope suspected that it was a piglike animal, but the fossils from out west established that the two teeth—a premolar and a molar from the left side of the mouth—belonged to a family of extinct, large, piglike creatures, now called entelodonts. Entelodonts were fearsome-looking animals, whose appearance roughly resembled a giant warthog. Their large canine teeth suggested to some paleontologists the possibility that they may have been at least occasional meat-eaters. Certainly the prospect of encountering a pack of wild entelodonts, snorting and grunting their way through the Miocene forest, is enough to make us thankful that these creatures are extinct.

Other animals from this deposit, the so-called Squankum local fauna, are also known largely from their teeth. These include teeth of a prehistoric semi-aquatic rhinoceros *Diceratherium* that had paired nasal horns; a peccary similar to the modern javelina of the American southwest; a deerlike member of the camel family, known as *Prosynthetoceras,* possessing branching nasal horns; a single primitive horse tooth (*Anchitherium*) from a three-toed, leaf-browsing horse; and a primitive carnivore tooth, perhaps from an ancestor of the dog family. Some of these same animals have been found in the lower part of the Calvert Cliffs exposures, in the Calvert Formation of late Early Miocene age. They are also similar to the fossil fauna found in the Miocene deposits of the Gulf Coast, especially to the early Miocene animals of east Texas.

All of these animals taken together suggest that the early Miocene was a time of warm climate and higher sea level. But as the Miocene progressed, the evidence indicates that the earth's climate became cooler and dryer. Grasslands replaced forests as the woodland retreated before the plains. It is strongly suspected that this ecological change was associated with the beginning of glacial conditions in the Antarctic continent. Soon the ice would spread further across the land's surface.

There are no good records of Pliocene fossils in New Jersey. There is a suspected Pliocene horse tooth from the stream-bed gravels of one of the Monmouth Brooks, but its precise stratigraphic origin is unknown. It is now thought that some of the yellowish gravels topping some of the low hills in the coastal plain, the Pennsauken Formation for instance, may be Pliocene in age, but in the absence of good contained fossils, precise correlations are lacking. The Pliocene was a time of generally arid climate, cooling temperatures, and lower sea level.

With the onset of the Pleistocene Epoch, the earth was plunged into a series of colder climate phases that have become popularly known as the Ice Age. In fact, over the last two million years, there were a number of colder intervals interspersed with times when the planet's climate was as warm or even slightly warmer than today. The classic sequence of glacial geology for the Pleistocene recognized four well-defined intervals of extensive glaciation that were called (from oldest to youngest) the Nebraskan, Kansan, Illinoisan, and Wisconsin glaciations. The intervening interglacial phases were named the Yarmouth, Aftonian, and Sangamon interglacials. Today some scientists recognize twenty or more cold spells with warmer phases in between.

Various theories have been promulgated to account for this climatic fluctuation. One old idea is that creation of widespread mountainous regions in the Miocene disrupted the ancient weather patterns and gradually cooled off the earth's climate. The Himalayas had begun to rise by then, squeezed upward in the collision between India and Asia. The Rockies and Alps were rising, and these highland regions provided areas for the birth of large glaciers. This idea in and of itself does not seem to be the whole cause, for these mountainous regions exist today, when we are in an interglacial phase. Another suggestion looks to the influence of continental drift. This hypothesis states that ice ages are favored when continents are positioned over or near the polar regions, as they are today.

Both of these suggestions seem more like preconditions for the steady pulsing of ice ages seen in the geologic record of the last two million years. Some more cyclical cause has been sought, and the answer may lie externally, in the slow but regular variations in the tilt and orbit of the earth that have come to be called the Milankovitch cycles, after the Serbian astronomer who discovered them.

Milankovitch demonstrated mathematically that the tilt of the earth's axis, which is now at 23.5 degrees to the plane of the earth's orbit around the sun, can vary between 21.5 degrees and 24.5 degrees. This tilt affects the concentration of sunlight reaching the earth's polar regions, and hence the amount of heat they receive. The tilt varies back and forth on a cycle that averages 41,000 years. This is combined with a shorter cycle of 23,000 years for what is known as axial precession, the regular variation in the direction of the axial tilt. Right now the northern hemisphere of the earth is tilted toward the sun in the summer, but this is slowly changing. A third Milankovitch cycle involves changes in the shape of the earth's orbit around the sun. The earth's orbit changes from more nearly circular, when the planet is generally slightly closer to the sun, to a more elliptical path when the earth can move slightly further away from the sun. It does so on a cycle of about 100,000 years. These orbital and tilt variations can all affect the amount and concentration of sunlight reaching the planet's surface, and therefore affect the earth's weather. The results may be the pattern of relatively rapid (geologically speaking) climate fluctuations resulting in the ice ages.

Other causes have been invoked as well, and it may be that the climate fluctuations of the Pleistocene are the result of several factors working together. There is some evidence from drill cores taken from the Greenland ice sheet that the earth's atmosphere has undergone changes in its composition which have helped precipitate ice ages, perhaps changes in carbon dioxide content, or sunlight transmissibility due to widespread volcanic activity. Other factors may include changes in oceanographic circulation patterns which bring moisture onto the land, and the effect of ice surges, giant slabs of ice sliding off the Antarctic ice sheet into the ocean.

Whatever the cause or combination of causes, abundant evidence exists to show that continental glaciation was once much more extensive than it is today. Modern ice sheets are restricted to Antarctica and Greenland, but at the time of the maximum advance of the great Pleistocene glaciers, one-third of the earth's land surface was covered by sheets of ice up to two miles in thickness. This area of glaciation included much of northern North America, including northern New Jersey. At the maximum extent of the Wisconsin ice sheet, about 18,000 years ago, the ice sheet ended in a long ridge of glacial debris derived from the melting of the glacier. That ridge of glacial debris can be traced across New Jersey from Belvidere on the Delaware River along a snaky twisting line to the vicinity of northern Raritan Bay. South of that wall of ice New Jersey was a wind-swept tundra where great prehistoric beasts of the Ice Age trudged in herds or hunted in solitary hunger.

In between the glacial advances, the climate warmed, as the mixture of some Pleistocene fossil faunas reveals. In a single locality we find evidence of some organisms that preferred cold weather and some that were better adapted to warmer weather. One of the best records of the Ice Age animals in the Mid-Atlantic region came from a famous fossil deposit that has since been buried. This is the Port Kennedy cave deposit in Valley Forge, Pennsylvania. Quarrymen working a layer of Paleozoic limestone in the late nineteenth century near the historic wintering grounds of Washington's army noticed a fissure filling in the carbonate rock. Limestone often develops caves, fissures, and sinkholes as it is dissolved by the percolation of slightly acid groundwater; this kind of landscape is called karstic, and it is responsible for the rolling topography of Valley Forge. One such sinkhole opened and was filled in during the Pleistocene; the sedimentary fill contained the remains of a variety of Ice Age animals and plants.[4] A number of insect specimens were recovered, mostly parts of ground beetles and dung beetles. Reptiles include fossil turtles and a snake, and even a specimen of a turkey was preserved.

The Port Kennedy mammals are a combination of the strange and the familiar, representing both cooler weather forms and animals from warmer climes. They are an assortment of rodents, mostly well-known forms such as squirrels, groundhogs, porcupines, beavers, muskrats, voles, and mice. The eastern cottontail is present, but so are specimens of the pika, a short-eared relative of the rabbit that inhabits cold mountainous regions today. Carnivores of various sorts were found in the fill, including many familiar types like otters, skunks, black bears, bobcats, foxes, and coyotes.

But there are also some more exotic predators present in the cave's fauna, and these can be broken down into a more southerly group and a northern faction. The jaguar, jaguarundi, and short-faced bear represent a warmer faunal influence, while the wolverine is a more northerly animal. There are also remains of a saber-toothed cat, *Smilodon gracilis,* and a cheetahlike cat. Among the larger mammalian herbivores, a familiar element is the white-tailed deer. But again a distinct southerly influence can be detected in the form of peccaries, tapirs, and the bones of the giant ground sloth *Megalonyx wheatleyi,* which were the most numerous of the larger mammal remains found in the deposit. On the other hand, the American mastodon, also found at this site, is usually interpreted as a cooler weather animal. Taken in its totality, the Port Kennedy fauna probably represents an averaged sample of animal populations from both cooler phases and warmer times during the middle of the Pleistocene.

In New Jersey, the most common of the large Pleistocene animals was the American mastodon, *Mammut americanum.* While this elephant relative had a trunk and tusks, it was also covered in shaggy hair. A major difference between true elephants and the mastodon is in the form of the teeth; elephants have relatively flat teeth, but mastodons have teeth with more pronounced conical cusps. Complete or nearly complete specimens of the American mastodon have been excavated from a number of sites in New Jersey. Many of the mastodon specimens were discovered in peat deposits, which formed in Pleistocene bogs (figure 10.5). One explanation for this association, most often seen in the mastodon discoveries in Warren and Sussex Counties, is that the great elephants became mired in the mucky bogs and were trapped there, starving to death and then sinking into the peat to be preserved in their entirety. This is similar to the scenario proposed for some of the other famous Ice Age fossil deposits such as the La

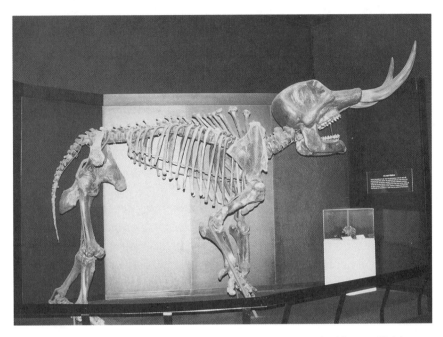

FIGURE 10.5 Mastodon skeleton. This specimen was excavated from a Pleistocene peat deposit in Liberty Township, Warren County, N.J. New Jersey State Museum.

Brea tarpits in Los Angeles, or for some of the frozen mammoths found preserved with soft parts intact in the Siberian tundra.

Mammoth specimens are also known from New Jersey, but they are much less frequent than mastodon fossils. Mammoths are also elephant relatives, some of which grew bigger than the modern elephants; they possessed large tusks that curved inward. Back in 1932 the partial skeletons of two mammoths were found during excavations for a golf course in Blackwood, Camden County, New Jersey. Barnum Brown of the American Museum of Natural History in New York was called in and he confirmed the identification. The discovery received much attention in the local press, but the bones themselves have since disappeared. Perhaps they will be rediscovered some day in an attic or in a museum collection somewhere.

More frequently, individual teeth of the Ice Age elephants are found. One such tooth was found in Trenton during excavations into the Trenton gravels along the Delaware River. There were several species of mammoths present in North America during the Pleistocene, including the imperial mammoth of the southwest, the Columbian mammoth of the southeast, and the wooly mammoth of the north. All of these species were more closely related to the living elephants of Africa and Asia than they were to the mastodon, which represents a separate and now extinct branch of elephant evolution. The Trenton tooth has been assigned to the wooly mammoth, a hairy large-tusked form that had a pronounced hump on the back.

Perhaps the most characteristically "Jerseyan" of all the Pleistocene megafauna is the elk-moose, *Cervalces scotti,* a cold-weather animal. The first specimen of this animal was dug up in 1887 in Mount Hope, Warren County.[5] It was described by the famous mammal paleontologist William Berryman Scott of Princeton University. The antlers are a cross between those of an elk and the palmate antlers of a moose. Specimens of this large relative of the deer have been found at a number of localities across North America since then, but the original skeleton remains the finest of its kind. It may still be seen mounted in the Natural History Museum in Guyot Hall on the campus of Princeton University. Another substantial portion of a *Cervalces* skeleton was found near Blairstown, Warren County, in 1969.

Closely coincident with the range of the elk-moose was the distribution of the giant beaver, *Castoroides ohioensis.* This aquatic rodent stood up to seven feet tall. A few specimens of its long gnawing teeth have been found in the Monmouth Brooks stream gravels; besides producing so many Cretaceous fossils, the stream-bed gravels also occasionally yield remains of Pleistocene megafauna derived from Pleistocene deposits eroded along the streams' courses. Other Pleistocene fossils found this way include the claw of a giant ground sloth, *Cervalces* bones and antlers, and mastodon remains.

Among the most anomalous Pleistocene fossils from our area are the bones and teeth of Ice Age land mammals dredged up from far off the coast by clammers and scallopers working the mollusk beds on the continental shelf east of New Jersey. The shell fishermen run their large steel dredges over the sea bottom, gouging up edible bivalves and the occasional fossil bone or tooth. The most frequent remains found this way are the tusks and skulls of walruses, which bespeak a colder climate off the coast of New Jersey not so long ago.

But the most interesting finds dredged up from Davy Jones's locker are

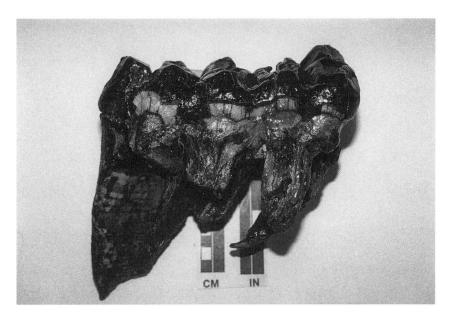

FIGURE 10.6 Mastodon tooth. Side view of tooth showing cusps at top and roots at bottom. Dredged up from the continental shelf off New Jersey. Donated by Captain John Larson. New Jersey State Museum

the remains of the large land mammals of the Pleistocene. Of these, the most frequently encountered are the teeth of mastodons and mammoths (see figures 10.6 and 10.7). Other land-mammal fossils found at some depth on the continental shelf include tapir, horse, caribou, musk ox, bison, *Cervalces,* and giant ground sloth.[6] These specimens are often found at depths of a hundred feet or more below the sea's present surface, at distances dozens of miles beyond the current shoreline. One unusual occurrence recorded from off Cape May was the recovery of the distinctively fused radius and ulna of a manatee. This specimen lay misidentified as a walrus in the collections of the Academy of Natural Sciences until correctly diagnosed by Barbara Grandstaff of Temple University. This find represents a very northerly record of these large peaceful plant-eating aquatic mammals, which are usually found today in the warm waters of the Gulf of Mexico and in the inland springs of Florida. Manatees are so temperature-sensitive that they will cluster around the warm water emanating from power plant outfalls

FIGURE 10.7 Mammoth tooth. Side view of tooth showing enamel ridges in tooth; flat surface on right is the grinding surface of the tooth. Dredged up from the continental shelf off New Jersey. Donated by Captain John Larson. New Jersey State Museum.

during cold spells in Florida. Recently (1995) manatees have been sighted in New Jersey waters during the summer months.

How did large land-mammal bones come to be dredged up from deep waters off the coast? A continental ice sheet one to two miles thick and covering many thousands of square miles is a very large ice cube. To make a glacier of this size requires a vast amount of water. That much water can only have come out of the ocean basins, by some change in oceanic circulation patterns coupled with a cooling of climate. As maritime snows piled up in places like the Laurentide Mountains of Canada, giant glaciers were generated that started moving south. The more the surface of the planet was covered with cold white glacial ice, the cooler the climate got, and less and less of the snow falling every year would melt. All of this moisture sequestered on the land drew down the level of the ocean, and large expanses of the continental shelf were exposed as dry land. At glacial maximum, sea level was three hundred to four hundred feet lower than today, and the shoreline of New Jersey was over a hundred miles east of the present coast. The great giants of the Pleistocene could wander out over what is now submerged sea floor, seeking plant food where clam banks flourish today. Since then the great ice sheets have melted, releasing their waters back into the ocean and flooding what was once land. This process continues as modern glaciers melt and sea level gradually rises.

What happened to the great mammals of the Ice Age? Obviously they are no longer here, and it has long been noted that the New World is relatively impoverished in large land animals when compared to the richer faunas of Africa and Asia. Thomas Jefferson engaged in a debate on this subject with the learned savants of Paris when he was ambassador to France, and when he became president he charged Lewis and Clark with the task of looking for live mastodons while they were on their famous expedition to explore the Louisiana Purchase.[7]

The traditional explanation for the extinction of the Pleistocene megafauna is that the big animals of the Ice Age could not survive the change of climate from the colder glacial phase to our present warmer interglacial phase. The problem with this is that most of these same animals managed to survive numerous previous changes in climate during the course of the Pleistocene. Some additional factor was involved in the megafauna's disappearance. Paul Martin, a paleontologist with the University of Arizona, has proposed that the additional factor was the arrival of humans on the North American continent.[8] He suggests that the North American Pleistocene mammals were unused to the presence of human beings and had no fear of this newcomer. We have seen this happen on some islands in modern times, where the first explorers remarked on the animals' lack of fear toward human beings, who then proceeded to wipe out entire species. Martin notes the evidence for mass killing techniques, such as firedrives and cliff stampedes, in which large numbers of animals could be killed with relatively little risk to the hunters. He sees a closely synchronized wave of extinction sweeping through North America around 11,000 to 10,000 years ago, coinciding with the spread of Paleoindian populations across the continent. Indeed, David Parris of the New Jersey State Museum has compiled radiocarbon (carbon-14) dates on mastodon specimens from New Jersey, New York, and eastern Pennsylvania that cluster between 12,000 and 10,000 years ago. The youngest dates are around 10,000 years old.[9]

Again, it may be a coincidence of causes that was responsible for the extinctions. Anthony Stuart of Castle Museum in Great Britain has analyzed the patterns and timing of the Pleistocene extinctions in both Eurasia and North America. He concluded that the disappearance of the large land mammals of the Pleistocene was due to the effects of human predation during a time of rapid environmental change.[10] This process continues today as species populations dwindle in the face of habitat destruction and environmental modification of the planet by humans. In the context of geologic time, we are still in the Pleistocene mass extinction episode as global biodiversity continues to decline and large prominent animals disappear into the fossil record.

A P P E N D I X   A

Where to See Dinosaurs and Other Fossils in and around New Jersey

Academy of Natural Sciences, 1900 Benjamin Franklin Parkway, Philadelphia, Pa. 19103. (215) 299-1000. "Discovering Dinosaurs" is an interesting exhibit hall that combines modern knowledge of dinosaur science with a history of ideas about dinosaurs. Includes a display of a reconstruction of a *Hadrosaurus foulkii* skeleton, a *T. rex* skeleton, horned dinosaur skeletons, a mosasaur and a plesiosaur skeleton, plus lots of information on the Cope-Marsh bone wars. Open seven days a week; admission charge. School programs are available.

American Museum of Natural History, 79th St. and Central Park, New York, N.Y. 10024-5192. (212) 769-5100. The newly renovated dinosaur halls contain the latest information on dinosaurs. Not to be missed by any true dinosaur aficionado. The new displays are thematically organized around the evolution and classification of dinosaurs using the methodology of cladistic analysis. Skeletons on display include *T. rex, Albertosaurus, Allosaurus, Apatosaurus, Barosaurus, Stegosaurus, Corythosaurus, Anatotitan, Triceratops,* and many other dinosaur specimens. Open seven days a week; admission charge.

Bergen Museum of Art and Science, 327 East Ridgewood Avenue, Paramus, N.J. 07652. (201) 265-1248. The remains of two mastodons are on display. Open Tuesday through Saturday.

Great Swamp Outdoor Education Center, Great Swamp National Wildlife Refuge, 247 Southern Boulevard, Chatham Township, N.J. 07928. (201) 635-6629. Has a display of dinosaur footprints, and information about local mastodon discoveries. Open daily, admission free.

Meadowlands Museum, 91 Crane Ave., Rutherford, N.J. 07070. (201) 935-1175. Has a small display of local fossils. Open Monday, Wednesday, and Sunday, in the afternoon; otherwise by appointment.

Morris Museum of Arts and Sciences, 6 Normandy Heights Road, Morristown, N.J. 07960. (201) 538-0454. Has a few exhibits about dinosaurs and fossils, including life-sized replicas of a stegosaur and a pterosaur, and some dinosaur footprints from nearby Florham Park. Open Tuesday through Sunday; admission charge.

New Jersey State Museum, 205 W. State St., CN-530, Trenton, N.J. 08625. (609) 292-6308. Exhibit in Natural History Hall focuses on New Jersey fossils and dinosaurs. Displays include a reconstruction of a hadrosaur skeleton, a horned dinosaur skeleton, dinosaur footprints, a skull of *Mosasaurus maximus,* a working fossil preparation lab, two mastodon skeletons, and a petrified tree trunk from New Jersey. Admission is free; closed on Mondays. Special school programs are available on request.

Newark Museum, 49 Washington St., Newark, N.J. 07101. (201) 596-6550. Has a few displays about paleontology; planned renovations include a mastodon skeleton. Open Wednesday through Sunday; admission charge.

Princeton Geology Museum, Guyot Hall, Princeton University, Princeton, N.J. 08540. Has a classic mounted skeleton of *Allosaurus,* a large late Jurassic theropod from Utah; a large *Diplodocus* thigh bone; a *Tyrannosaurus rex* skull; dinosaur babies and eggs from Montana; a *Coelophysis* skeleton; the original *Cervalces* (elk-moose) skeleton; and a wide range of fossil displays. Admission is free; open seven days a week.

Rutgers Geology Museum, Geology Hall (2nd Floor), George and Somerset Streets, College Avenue Campus, New Brunswick, N.J. 08903. (908) 932-7243. Also concentrates on local paleontology and geology, with a nice slab of dinosaur trackways from the Newark redbeds and an exhibit on New Jersey hadrosaurs. A partial sauropod skeleton is on display, as well as a mosasaur skull, a mastodon skeleton, a Cretaceous theropod footprint, and a model of a small early bipedal dinosaur. Admission is free; guided tours for school groups are available by appointment. Open Monday through Friday.

Trailside Nature and Science Center, Coles Avenue and New Providence Road, Mountainside, N.J. 07092. (908) 789-3670. Life-size dinosaur and flying reptile replicas plus fossil footprints and local fossil specimens on display. Open daily, admission free.

Wagner Free Institute of Science, 1700 West Montgomery Ave., Philadelphia, Pa. 19121. (215) 763-6529. A veritable museum of a museum, the Wagner has authentically preserved a true Victorian museum. Nothing has changed since Cope and Leidy lectured here, and a partial skeleton of a sauropod that Cope obtained from Colorado is on display upstairs. There are numerous other fossils on exhibit, arranged in typical nineteenth-century fashion. Admission is free; open Tuesday through Friday. The Wagner also sponsors free classes in geology and paleontology, and conducts school programs and guided tours by appointment.

A P P E N D I X B

Methods for Studying Dinosaur Footprints

In some sections of the Newark Basin deposits, dinosaur footprints are relatively common. As noted in Chapter 5, the redbeds of Late Triassic and Early Jurassic age that occupy the north-central part of New Jersey contain the record of the earlier part of the age of dinosaurs. Here, excavations and outcrops may reveal the trackways of some of the smaller early dinosaurs. There are several methods for studying and collecting dinosaur footprints; it is not always practical to physically remove them and take them intact back to the lab or museum.

COLLECTING

The most preferable way to study dinosaur footprints is to quarry them out of the rock and transport them back to the laboratory, school, or museum. This is not always possible for several reasons. The tracks may be located on private property or publicly protected land where collecting is not permitted. One should always obtain permission from the property owners or other appropriate authorities before collecting fossil specimens. In some cases, such as construction projects and excavations, piles of rock rubble slated for disposal or burial may yield footprints, and this material should be easily removable. But in other cases trackways may lay embedded in fragile or very tough bedrock, and removal of the footprints may become impossible without destroying them. Such is the case with the layers that contain the actual negative or molds of the original footprints. Frequently the print-bearing layer is composed of very weak, thin-bedded, easily broken shale or mudstone; when struck with a hammer or chisel, the rock will tend to break at its weakest point which is usually the thinner area of the footprint. Often the overlying layer is a stronger, more thickly bedded siltstone or sandstone that slabs out very nicely, preserving the natural cast or positive of the footprint, the raised surface of the sediment that filled in the original negative of the track. Sometimes a large footprint, such as the type known as *Eubrontes,* may be present on the surface of a big thick slab of tough siltstone; attempting removal of this kind of specimen can be very difficult because of the sheer mass involved. If footprints are cracked upon extraction, the pieces may be reassembled more easily back in the lab with the aid of "witness marks," arrows written in indelible ink on the broken pieces of footprint slab indicating how the pieces fit back together. It may also be desirable to draw a directional arrow on the slab, pointing to north, if the footprint is removed from outcrop. Other data may be noted as well directly on the slab of rock.

Transporting and storing lots of slabs of large heavy rock can become a major logistical challenge even for experts. One major museum (the Yale Peabody Museum) has solved this problem by incorporating dinosaur footprint slabs into the walls of its collection storage area, but others may wish to consider alternative ways to "collect" dinosaur footprints.

CASTING

The most popular alternative to collecting the actual tracks is to make a plaster or latex cast of them. These may be made in the field or the lab. For plaster casting, a simple casting frame can be made out of furring strips or other lumber. For small prints, even a large aluminum pie dish will do. An adjustable frame can be constructed by nailing together four pieces of lumber to form two L-shaped sections, which can be secured together by thick rubber bands; this can be adjusted to any desired length and placed over the footprint. The footprint and surrounding area should be smeared thoroughly with a lubricant such as petroleum jelly to assure that the hardened plaster can be easily lifted from the rock surface when the job is done. Then the plaster is mixed and poured into the frame and allowed to set. The result should be an accurate reproduction of the footprint, which can be painted a suitable color to enhance its similarity to the real track.

Plaster is widely available and relatively inexpensive, but latex enjoys several advantages over plaster. It is lighter in weight, more flexible, sets more rapidly, produces a more finely detailed cast, and is easier to transport and store. The major drawbacks of latex are that it tears easily, and can be degraded and weakened by prolonged exposure to light.

PHOTOGRAPHY

Sometimes the best way to "collect" footprints (and other fossils as well) is to take pictures. This can provide an accurate record of the field associations of the tracks to other footprints and to the rocks in which they are found. Photographs of faint or obscure tracks may be enhanced by outlining the footprints in chalk, or by wetting. Low-angle lighting will increase shadowing, emphasizing the footprint.

MAPPING

It is often advisable to draw a detailed field map of a bedding-plane surface that has dinosaur footprints on it. This can preserve useful information about the relationships of the tracks to each other and to sedimentary features. Compass orientations, bedding dip, dimensions of tracks, and distance between footprints are some of the data that should be recorded. Such maps can be used to delineate trackways of individual animals, for information about footprint density and diversity, to determinate preferred directions if present, and to calculate speed based on stride length. Analysis of trackway maps is aided by scanning maps or footprint images into computers for statistical treatment.

RUBBINGS

A quick but effective way to make reproductions of footprints is by producing rubbings. Since dinosaur footprints are essentially two-dimensional, this can be easily done in mass quantity. Simply take a sheet of paper and hold it against the footprint while rubbing the paper's surface with the side of a sharpened pencil. An accurate copy of the track in the form of dark shading on the paper is the end product. This is an economical and clean way to make footprint images, and it seems especially suited to school classes or other educational groups.

DINOSAUR FOOTPRINT SITES

Dinosaur footprints have been found at a number of localities in New Jersey, but they are more common higher up in the stratigraphic stack of the Newark Supergroup rocks along the northwestern part of the Newark Basin. Dinosaur tracks are known from Newark, Little Falls, Roseland, New Vernon, Pompton Falls, Millington, Avondale, Florham Park, Whitehall, Frenchtown, and Milford. Across the state borders, the Newark Supergroup rocks have produced dinosaur footprints in the redbeds of southeastern Pennsylvania and in the Newark sediments of New York along the Hudson River north of the Palisades. No dinosaur bones have as yet been found in New Jersey's Newark Basin, although Newark Supergroup rocks in the Connecticut Valley of Massachusetts and along the Bay of Fundy in Nova Scotia have yielded rare and usually fragmentary dinosaur remains.

A P P E N D I X   C

How to Find Fossils in New Jersey

Many New Jersey fossil sites are located on private property, and permission to collect fossils on private land should always be sought from the property owner. On public lands, permission to collect should be obtained beforehand from the appropriate authorities responsible for that land. For instance, there are many fossil-bearing outcrops of Paleozoic rocks exposed around the mountains on either side of the Delaware River above the Delaware Water Gap. Much of this land is under the jurisdiction of the Delaware Water Gap National Recreation Area, and a permit from the National Park Service of the U.S. Department of the Interior is required to legally collect fossils from outcrops in the Delaware Water Gap National Recreation Area.

Frequently there are short-term opportunities to collect fossils arising from excavations or construction that temporarily expose fossiliferous sediments. Scouting such temporary exposures may prove rewarding, especially in areas where there are known fossil-bearing beds. Again, obtain permission from the people in charge before attempting to prospect construction or excavation sites.

Quarries, pits, and road cuts are also potential fossil sites. These sites are also potentially dangerous. Commonsense precautions should be observed whenever collecting takes place in steep-sided rock cuts or excavations of any sort. Be aware of unstable slopes above or below you, which may give way without warning. Cliff-falls are serious life-threatening events, and collectors should avoid undercutting steep slopes or hopping around on unstable cliff faces. Hardhats may be required for protection from rockfalls. Again, contact the authorities in charge before entering active mining operations.

One old quarry that is accessible to collectors, the Walter T. Kidde Memorial Dinosaur Footprint Quarry, is now part of a county park system. Permission to collect at this site can be obtained by contacting the Essex County Park Commission at (201) 268-3500. This locality has produced numerous dinosaur footprints of the type described and illustrated in Chapter 5 (see figures 5.7 through 5.11).

Particularly in the inner coastal plain region of southern New Jersey, streams and brooks expose outcrops of fossiliferous sediment in the steeper banks along their courses. Stream gravels may concentrate fossils eroded out of the banks. The best way to work the gravel is to use a sieve made out of wire mesh nailed to the bottom of a rectangular wooden frame. Simply shovel some stream gravel into the sieve and then wash away the sand and mud; a fossil-rich concentrate will be left if one is near an outcrop that yields fossils. One place where such a stream-cut fossil site is available to collectors is Poricy Brook Fossil Park in Middletown, New Jersey. Besides fossil seashells and sharks' teeth such as those depicted in figures 7.5, 7.6, and 7.7, one may also find dinosaur teeth or bones in the stream-bed gravels where they cut through deposits of Cretaceous age.

Elsewhere, while the underlying sediments may be barren of fossils, surface gravels underlying higher ground often contain fossiliferous pebbles derived by erosion and transport from the Paleozoic rocks in the Appalachian region. A quick inspection of chert pebbles contained in the yellowish or orange gravels that cap many southern New Jersey hills will reveal an abundance of Paleozoic sea animal fossils of the types illustrated in figure 2.3. Many stream-bed gravels in the little rivers of the Pinelands region will yield the same kinds of Paleozoic fossils, and these same fossil chert pebbles can also be found along the beaches of the Atlantic shore and on Delaware Bay at Sunset Beach.

Occasionally, fossil shark teeth may be found along the beaches of the New Jersey shore, especially around inlets. Places where this has occurred are Shark River Inlet, Corson's Inlet, and Townsends Inlet. Typically the teeth are Tertiary forms such as those illustrated in figure 10.4. Very rarely, remains of Ice Age mammals may wash up on the beach, particularly after a storm. Back in the nineteenth century, the ankle bone of a giant ground sloth was found in the surf at Point Pleasant. More recently, a mastodon tooth was found at Island Beach State Park, and a mammoth tooth was discovered on a sand bar at Holgate on Long Beach Island. These are spotty occurrences, but something to be aware of when beachcombing during a visit to the shore.

The kinds of tools and equipment needed for collecting fossils will vary depending upon the type of site being worked. Hard rock sites such as those in the northern part of the state will require rock hammers, pry bars, chisels, and protective eyeware such as goggles. In the southern coastal plain, the sediments are for the most part unconsolidated, and entrenching tools, shovels, trowels, and sieving boxes will be more helpful. In both areas, collecting bags, aluminum foil, glue, and small plastic vials (such as film canisters) are likely to be useful for storing and repairing specimens. Masking tape can be written on to label specimens or bags of specimens in the field; a field notebook is also very useful for recording detailed information about the fossils you have found.

Dinosaur fossils are rare, but they may be found in many of the above-mentioned circumstances if Mesozoic deposits are present. If you do find large heavy bones in the ground, it is usually a good idea to contact a museum with experienced paleontologists on its staff. They will be able to tell you if the bones are indeed those of a fossil animal from long ago. Removing fossil bones is a delicate and painstaking task; although they may appear hard, the bones can be easily broken or destroyed if they are not handled properly. Moreover, valuable scientific information may be gathered by careful excavation. For these reasons, if you think you have found a fossil skull or even a whole skeleton, please call the New Jersey State Museum or another local museum with a paleontology department if you need assistance.

Notes

CHAPTER 1

1. A. J. Desmond, *The Hot-Blooded Dinosaurs: A Revolution in Paleontology* (New York: Dial Press, 1975), 7–11.
2. H. S. Torrens, The dinosaurs and dinomania over 150 years, *Modern Geology,* 18 (1993): 270–275.
3. Torrens, The dinosaurs and dinomania, 277–281; also Desmond, *Hot-Blooded Dinosaurs,* 19–24.

CHAPTER 2

1. B. F. Howell, *Revision of the Upper Cambrian Faunas of New Jersey,* Geological Society of America Memoir 12 (1945), 1–46.
2. For more detailed information on Paleozoic fossil sites in New Jersey see W. B. Gallagher, Paleoecology of the Delaware Valley region, Part 1: Cambrian to Jurassic, *The Mosasaur,* 1 (1983): 23–40.

CHAPTER 3

1. E. Daeschler, N. Shubin, K. Thomson, and W. W. Amaral, A Devonian tetrapod from North America, *Science,* 265 (1994): 639–642.
2. P. Sereno, C. A. Forster, R. R. Rodgers, and A. M. Monetta, Primitive dinosaur skeleton from Argentina and the early evolution of Dinosauria, *Nature,* 361 (1993): 64–66.

CHAPTER 4

1. H. S. Torrens, The dinosaurs and dinomania over 150 years, *Modern Geology,* 18 (1993): 263–275.
2. D. Baird, The dinosaur industry in Philadelphia, 1787–1876, in *Proceedings of the Cope Symposium* (Philadelphia, PA: Wagner Free Institute of Science, in preparation).
3. R. W. Howard, *The Dawnseekers: The First History of American Paleontology* (New York: Harcourt, Brace, Jovanovich, 1975), 57.
4. G. Montgomery, New Jersey's Haddonfield dinosaur: A surprising history, *New Jersey Outdoors,* 11 (1984): 8–9, 30–31.
5. W. P. Foulke (Discovery of *Hadrosaurus foulkii*), *Proceedings of the Academy of Natural Sciences of Philadelphia,* 10 (1858): 211–214.
6. J. Leidy (Remarks on *Hadrosaurus foulkii*), *Proceedings of the Academy of Natural Sciences of Philadelphia,* 10 (1858): 215–218.
7. R. C. Ryder, Dusting off America's first dinosaur, *American Heritage,* 39 (1988): 68–73.

8. A. Desmond, *Archetypes and Ancestors: Palaeontology in Victorian London* (Chicago: University of Chicago Press, 1982), 129.

9. J. Leidy, *Cretaceous Reptiles of the United States,* Smithsonian Contributions to Knowledge, no. 192 (1865), Washington, D.C., 135 pp. plus XX plates.

10. T. B. Holmes and M. W. Pharo, *The Haddonfield Home of Edward Drinker Cope* (Haddonfield, N.J.: Haddonfield Historical Society, 1992), 1–25.

11. H. F. Osborn, *Cope: Master Naturalist* (Princeton: Princeton University Press, 1931), 156–176.

12. E. D. Cope, (Remarks on *Laelaps aquilunguis*), *Proceedings of the Academy of Natural Sciences of Philadelphia,* 18 (1866): 275–279.

13. Osborn, *Cope,* 468.

14. J. H. Ostrom, *Osteology of* Deinonychus antirrhopus, *an unusual theropod from the Lower Cretaceous rocks of Montana,* Bulletin 30 (1969), Peabody Museum of Natural History, Yale University, New Haven, Conn., 165 pp.

CHAPTER 5

1. M. J. Benton, Late Triassic extinctions and the origin of dinosaurs, *Science* 260 (1993): 769–770.

2. G. Jepsen, A natural library, New Jersey State Museum Bulletin 3 (1949), Trenton, N.J., 6 pp.

3. D. Baird, Some Upper Triassic reptiles, footprints, and an amphibian from New Jersey, *The Mosasaur* 3 (1989): 142–150.

4. B. Cornet, A. Traverse, and N. G. McDonald, Fossil spores, pollen, fishes from Connecticut indicate Early Jurassic age for part of the Newark Group, *Science* 182 (1973): 1243–1247.

5. P. E. Olsen and D. Baird, The ichnogenus *Atreipus* and its significance for Triassic biostratigraphy, in K. Padian, ed., *The Beginning of the Age of Dinosaurs* (New York: Cambridge University Press, 1986), 61–87.

6. P. E. Olsen, Paleontology and paleoenvironments of Early Jurassic age strata in the Walter Kidde Dinosaur Park (New Jersey, USA), in J.E.B. Baker, ed., *Contributions to the Paleontology of New Jersey,* Proceedings and Field Trips, Geological Association of New Jersey, William Patterson College, Wayne, N.J., Vol. XII (1995), 173–175.

7. W. B. Gallagher, Dinosaurs of the Newark Rift Basin, in J. M. Husch and M. J. Hozik, eds., *Geology of the Central Newark Basin,* Field Guide and Proceedings of the Fifth Annual Meeting of the Geological Society of New Jersey, Rider College, Lawrenceville, N.J. (1988), 215–230.

CHAPTER 6

1. J. Leidy (Remarks on *Antrodemus*), *Proceedings of the Academy of Natural Sciences of Philadelphia,* 22 (1870): 3–4.

2. D. Norman, *The Illustrated Encyclopedia of Dinosaurs* (New York: Crescent Books, 1985), 90–91.

3. P. M. KRANZ, *Dinosaurs in Maryland,* Maryland Geological Survey, Department of Natural Resources, Educational Series No. 6 (1989), 14.

4. J. D. McLENNAN, *Dinosaurs in Maryland,* Maryland Geological Survey, Department of Natural Resources (unnumbered pamplet, 1973).

5. D. GILETTE, True grit, *Natural History,* 104 (1995): 41–43.

6. R. BAKKER, *The Dinosaur Heresies* (New York: William Morrow and Company, Inc., 1986), 179–198.

7. C. D. MICHENER AND D. A. GRIMALDI, A *Trigonia* from Late Cretaceous amber of New Jersey (Hymenoptera: Apidae: Meliponinae), *American Museum Novitates,* Number 2917 (1988): 1–10.

8. E. O. WILSON, F. M. CARPENTER, AND W. L. BROWN, The first Mesozoic ants, *Science,* 157 (1967): 1038–1040.

9. L. F. GALL AND B. H. TIFFNEY, A fossil noctuid moth egg from the Late Cretaceous of Eastern North America. *Science,* 219 (1983): 507–509.

10. R. DESALLE, J. Gatesy, W. Wheeler, and D. Grimaldi, DNA sequences from a fossil termite in Oligo-Miocene amber and their phylogenetic implications, *Science,* 257 (1992): 1933–1036.

11. D. BAIRD, Medial Cretaceous dinosaur and footprints from New Jersey, *The Mosasaur,* 4 (1989): 53–63.

12. BAIRD, Medial Cretaceous dinosaur, p. 62.

CHAPTER 7

1. For more information about Cretaceous fossils in New Jersey, see W. B. Gallagher, Paleoecology of the Delaware Valley region, Part II: Cretaceous to Quaternary, *The Mosasaur,* 2 (1984): 9–43.

2. S. WELLER, *A Report on the Cretaceous Paleontology of New Jersey,* New Jersey Geological Survey, Paleontological Series Vol. 4 (1907), Trenton, N.J.

3. D. BAIRD AND J. R. HORNER, Cretaceous dinosaurs of North Carolina, *Brimleyana,* 2 (1979): 1–28.

4. R. M. ALEXANDER, *Dynamics of Dinosaurs and Other Extinct Giants* (New York: Columbia University Press, 1989), 129–138.

5. D. C. PARRIS, Additional records of plesiosaurs from the Cretaceous of New Jersey, *Journal of Paleontology,* 48 (1974): 32–35.

6. D. A. RUSSELL, *Systematics and Morphology of American Mosasaurs,* Bulletin 23 (1967), Peabody Museum of Natural History, Yale University, New Haven, Conn., 60–69.

7. W. B. GALLAGHER, The Cretaceous/Tertiary mass extinction in the northern Atlantic Coastal Plain, *The Mosasaur,* 5 (1993): 75–154.

8. E. G. KAUFFMAN AND R. V. KESLING, An Upper Cretaceous ammonite bitten by a mosasaur, *Contributions from the Museum of Paleontology,* University of Michigan, 15 (1960): 193–248.

9. J. E. Martin and P. R. Bjork, Gastric residues associated with a mosasaur from the Late Cretaceous (Campanian) Pierre Shale in South Dakota, *Dakoterra*, 3 (1987): 68–72.

10. D. Baird and P. M. Galton, Pterosaur bones from the Upper Cretaceous of Delaware, *Journal of Vertebrate Paleontology*, 1 (1981): 67–71.

CHAPTER 8

1. J. Leidy, (Remarks on *Hadrosaurus foulkii*), *Proceedings of the Academy of Natural Sciences of Phildelphia*, 10 (1858): 215–218.

2. J. Horner, Upper Cretaceous Dinosaurs from the Bearpaw Shale (Marine) of south-central Montana with a checklist of Upper Cretaceous dinosaur remains from marine sediments in North America, *Journal of Paleontology*, 53 (1979): 566–577.

3. J. Horner, *Digging Dinosaurs* (New York: Workman Publishing, 1988), 210 pp.

4. J. Leidy, *Cretaceous Reptiles of the United States*, Smithsonian Contributions to Knowledge, no. 192 (1865), 98.

5. L. Woolman, Bone of a Dinosaur, an immense reptile, associated with ammonites and other molluscan fossils in Cretaceous (Matawan) Clay Marls at Merchantville, N.J., *Annual Report of New Jersey Geological Survey for 1896*, Trenton, N.J., 248–250.

6. Leidy, *Cretaceous Reptiles*, 119.

7. E. D. Cope, (Remarks on *Laelaps aquilunguis*), *Proceedings of the Academy of Natural Sciences of Philadelphia*, 18 (1866): 275–279; E. D. Cope, Synopsis of the extinct batrachia, reptilia and aves of North America, *Transactions of the American Philosophical Society*, 14 (1871–1872): 118–119.

8. Leidy, *Cretaceous Reptiles*, 102–103.

9. Cope, (Remarks on *Laelaps*), 275–279.

10. R. E. Molnar, Problematic Theropoda: "Carnosauria," in D. B. Weishampel, Peter Dodson, H. Osmolska, eds., *The Dinosauria* (Berkeley, Cal.: University of California Press, 1990), 310–311.

11. R. Bakker, *The Dinosaur Heresies* (New York: William Morrow and Company, Inc., 1986), 481 pp.

12. R. M. Alexander, *Dynamics of Dinosaurs and Other Extinct Giants* (New York: Columbia University Press, 1989), 27–59.

13. A. Chinsamy and P. Dodson, Inside a dinosaur bone, *American Scientist*, 83 (1995): 174–180; see also A. Chinsamy, L. M. Chiappe, and P. Dodson, Growth rings in Mesozoic birds, *Nature*, 368 (1994): 196–197.

14. O. C. Marsh, (Remarks on *Hadrosaurus minor*), *Proceedings of the Academy of Natural Sciences of Philadelphia*, 22 (1870): 2–3.

15. E. H. Colbert, A hadrosaurian dinosaur from New Jersey, *Proceedings of the Academy of Natural Sciences of Philadelphia*, vol. C (1948): 23–27.

16. D. BAIRD AND J. R. HORNER, A fresh look at the dinosaurs of New Jersey, *New Jersey Academy of Sciences Bulletin,* 22 (1977): 50. Also, W. B. Gallagher, *Dinosaurs: Creatures of Time,* New Jersey State Museum Bulletin No. 14 (1990), Trenton, N.J.

CHAPTER 9

1. S. L. OLSON AND D. C. PARRIS, The Cretaceous Birds of New Jersey, *Smithsonian Contributions to Paleobiology,* No. 63 (1987), Washington, D.C., 1–22.
2. B. S. GRANDSTAFF, D. C. PARRIS, R. C. DENTON, JR., AND W. B. GALLAGHER, *Alphadon* (Marsupalia) and Multituberculata (Allotheria) in the Cretaceous of eastern North America, *Journal of Vertebrate Paleontology,* 12 (1992): 217–222.
3. L. W. ALVAREZ, W. ALVAREZ, F. ASARO, AND H. MICHEL, Extraterrestrial cause for the Cretaceous-Tertiary extinction, *Science,* 208 (1980): 1095–1107.
4. L. T. SILVER AND P. H. SCHULTZ, eds., *Geological Implications of Impacts of Large Asteroids and Comets on the Earth,* Geological Society of America Special Paper 190 (1982), Boulder, Colorado, 528 pp. Also: V. I. Sharpton and P. Ward (eds), *Global Catastrophes and Earth History,* Geological Society of America Special Paper 247 (1990), Boulder, Colorado, 631 pp.
5. A. R. HILDEBRAND, G. T. Penfield, D. A. Kring, M. Pilkington, Z. A. Camargo, S. B. Jacobsen, and W. V. Boynton, Chicxulub crater: A possible Cretaceous-Tertiary Boundary impact crater on the Yucatan Peninsula, Mexico, *Geology,* 19 (1991): 867–871.
6. W. B. GALLAGHER, The Cretaceous/Tertiary mass extinction event in the northern Atlantic Coastal Plain, *The Mosasaur,* 5 (1993): 75–154.
7. W. B. GALLAGHER, Geochemical investigations of the Cretaceous-Tertiary boundary in the Inversand Pit, Gloucester County, New Jersey, *New Jersey Academy of Science Bulletin,* 37 (1992): 19–24.
8. W. B. GALLAGHER, Selective extinction and survival across the Cretaceous/Tertiary boundary in the northern Atlantic Coastal Plain, *Geology,* 19 (1991): 967–970.
9. S. L. PIMM, H. L. JONES, AND J. DIAMOND, On the risk of extinction, *American Naturalist,* 132 (1989): 757–785.
10. W. B. GALLAGHER, 1991, op. cit.
11. S. L. PIMM, et al., 1989, op. cit.

CHAPTER 10

1. E. D. COPE, Synopsis of the extinct batrachia, reptilia and aves of North America, *Transactions of the American Philosophical Society,* Vol. 14 (1871–1872): 228.
2. J. LEIDY, Notice of some remains of extinct pachyderms, *Proceedings of the Academy of Natural Sciences of Philadelphia,* 19 (1867): 230–233.
3. H. G. RICHARDS AND A. HARBISON, Miocene invertebrate fauna of New Jersey, *Proceedings of the Academy of Natural Sciences of Philadelphia,* 44 (1942): 167–250.

4. E. Daeschler, E. E. Spamer, and D. C. Parris, Review and new data on the Port Kennedy local fauna and flora (Late Irvingtonian), Valley Forge National Historical Park, Montgomery County, Pennsylvania, *The Mosasaur,* 5 (1993): 23–41.

5. D. C. Parris, New and revised records of Pleistocene mammals of New Jersey, *The Mosasaur,* 1 (1983): 1–21.

6. W. B. Gallagher, D. C. Parris, B. S. Grandstaff, and C. DeTample, Quaternary mammals from the continental shelf off New Jersey, *The Mosasaur,* 4 (1989): 101–110.

7. R. W. Howard, *The Dawnseekers: The First History of American Paleontology* (New York: Harcourt, Brace, Jovanovich, 1975), 34, 74.

8. P. Martin, Pleistocene overkill. In: P. Martin and H. E. Wright, eds., *Pleistocene Overkill: The Search for a Cause* (New Haven: Yale University Press, 1967), 75–120.

9. Parris, New and revised records, pp. 17–21; also, D. C. Parris, J. Camburn, and S. Camburn, A megafaunal radiocarbon date for a Monmouth County Pleistocene brook locality, *Journal of Vertebrate Paleontology,* Abstracts of Papers, 15 (1995): 47A–48A.

10. A. J. Stuart, Mammalian extinctions in the Late Pleistocene of Northern Eurasia and North America, *Biological Review,* 66 (1991): 453–562.

Glossary

acetabulum: hip socket into which fits the upper head or ball of the thigh bone (femur); in dinosaurs the acetabulum is perforated.

ammonites: extinct group of cephalopods with chambered shells possessing complexly crenelated chamber walls.

anapsids: members of the reptilian subclass Anapsida, characterized by skulls of solid bone without any temporal openings behind the eyes; in addition to various extinct groups of reptiles, this includes the turtles.

angiosperms: flowering plants.

ankylosaurs: armored dinosaurs, whose bodies were covered in bony plates.

archosaurs: literally, "ruling reptiles"; includes various primitive reptilian groups of thecodont grade, as well as crocodilians, pterosaurs, and dinosaurs. Characterized by an extra opening in the skull in front of the eyes.

argillite: a kind of hard shale or mudstone that has some salts in it; in New Jersey, it is the principal component of the Lockatong Formation.

astagalus: one of the bones of the ankle.

belemnites: extinct relatives of the squid that grew thick cylindrical carbonate shells, sometimes called squid pens.

brachiopod: primitive bivalved shellfish that were common in the Paleozoic seas.

bryozoa: colonial relatives of the brachiopods; major components of the Paleozoic reef community.

calcaneum: one of the bones of the ankle.

caliche: calcareous deposit found forming today in arid and semi-arid soils, in the American Southwest for example; also found in redbeds of the Newark Supergroup of New Jersey.

carbonate: rocks containing the carbonate ion (CO_3), such as limestone or dolomite. Usually laid down in shallow warm waters.

carnosaur: any of the larger, later meat-eating dinosaurs generally characterized by the reduction of the number of fingers on the hands. Includes forms like *Allosaurus* and *Tyrannosaurus.*

catastrophism: the view that some geologic phenomena (such as certain mass extinctions) are best explained as the result of large-scale and very rapid events.

cephalopods: literally, "head-foot"; members of the molluscan family that have tentacles attached to the head region. Includes squid, octopus, and the extinct ammonites and belemnites.

ceratopsians: horned dinosaurs of the order Ornithisichia. Found in Late Cretaceous deposits.

coccolithophorids: minute marine algae that secrete a shell composed of small aggregated carbonate disks. The dissociated disks are a principal component of the Cretaceous chalks.

coelacanth: lobe-finned fishes possessing a three-lobed tail; in addition to the coelacanths of the Late Paleozoic and Mesozoic, includes the living fossil *Latimera* found off the coast of East Africa.

coevolution: the idea that species of plants and animals evolve, or change, in response to each other's changes.

continental drift: the idea that the continents have changed position, moving very slowly over the surface of the globe over the course of geologic time.

Cretaceous: last period of the Mesozoic Era, lasting from 144 million to 65 million years ago; from the Latin word for chalk.

crinoids: members of the echinoderm phylum; they consist of a head or calyx positioned atop a segmented stalk. Also known as sea lilies because of their flowerlike appearance.

crossopterygians: lobe-finned fish, including many Paleozoic forms as well as the living fossil coelacanth.

cuesta: a low, continuous ridge usually found in coastal regions.

cycads: low-growing trees with palmlike foliage and seed-cone reproductive bodies; common in the Mesozoic Era.

deltopectoral crest: the bony flange on the upper part of the humerus (upper bone of the arm).

diapsid: a member of the reptilian subclass Diapsida, characterized by two openings in the temporal region of the skull, behind the eyes.

dinosaur: extinct archosaurian reptiles characterized by erect posture of the limbs and a perforated acetabulum.

elasmosaurs: long-necked plesiosaurs with small skulls; named after *Elasmosaurus* described by E. D. Cope in 1872.

entelodonts: extinct large relatives of the pigs, Oligocene to Miocene in age.

eurypterids: sea scorpions of the Early Paleozoic Era; relatives of the modern horseshoe crab, some eurypterids grew up to seven feet in length.

foraminifera: zooplankton that secretes a carbonate shell.

formation: a mappable unit of rock with distinctive physical characteristics such as color, composition, grain size, etc. Usually named for a geographic area where it is exposed, e.g., the Mount Laurel Formation.

gastroliths: stones swallowed by animals to aid in processing food; sometimes found associated with herbivorous dinosaur skeletons, especially sauropods.

gastropods: literally, "stomach-foot"; the snails of the phylum Mollusca.

gharial: thin-snouted crocodilians of southern Asia; also known as gavials.

glaciation: an interval of widespread development of large, thick glaciers; the condition of being covered by glaciers.

glauconite: a complex hydrated iron aluminum potassium silicate, in the mica family; green in color, it is the principal constituent of greensand. Deposited under marine conditions.

gradualism: the view that geologic changes can generally be ascribed to slow earthly processes; opposed to catastrophism.

graptolites: extinct colonial floating relatives of vertebrates; common in the oceans of the Early Paleozoic Era.

greensand marl: glauconitic deposits once extensively dug in the marl belt of inner coastal plain sediments extending across New Jersey from Raritan Bay in the northeast diagonally across the state to near Delaware Bay in the southwest.

gymnosperms: literally, "naked seed"; seed-bearing plants whose seeds are not encased in an outer covering, as is the case with the seeds of flowering plants. Includes conifers and cycads.

hadrosaurs: the family of Late Cretaceous duck-billed dinosaurs, named for *Hadrosaurus.*

heteromorph: a type of ammonite in which the shell is complexly coiled, sometimes coiling, uncoiling, and recoiling, as in *Anaklinoceras.*

hybodonts: extinct spiny sharks of the Late Paleozoic and Mesozoic Eras.

hyponome: muscular nozzle used for jet propulsion by cephalopods.

igneous: referring to rocks formed from the cooling and solidification of magma.

inoceramids: family of extinct clams that were common in the Mesozoic seas; some Cretaceous forms grew up to a yard in length.

interglacial: an interval between major glaciations that is usually warmer, during which sea level is generally higher than during glacial episodes.

Jurassic: the middle period of the Mesozoic Era, from 208 million years ago to 144 million years ago. Named after outcrops in the Jura Mountains on the border between Switzerland and France.

karstic: referring to the type of landscape characterized by caves, sinkholes, and underground streams and rivers; always developed on carbonate substrates.

lambeosaurine: referring to the subfamily of crested hadrosaurs possessing bony extensions of the nasal bones.

lamnoids: spineless sharks with streamlined bodies; includes most of the large predatory sharks of modern waters.

Liassic: referring to rocks of Early Jurassic age, usually marine shales, well exposed in England and Germany.

lignite: fossil wood in the first stages of conversion into coal.

marl: a loose clay or clayey sand containing calcium carbonate, often in the form of fossil shells. Mined for use as a soil dressing.

megafauna: usually used to refer to the large mammals of the Pleistocene Epoch, many of which disappeared about 11,000 to 10,000 years ago.

megalosaurian: pertaining to the mostly European group of theropod dinosaurs of Jurassic and Early Cretaceous age; named after *Megalosaurus,* the first carnivorous dinosaur known to science, described by William Buckland in 1824.

metamorphic: referring to rocks formed from previously existing rocks by recrystallization; in this process, heat and pressure change the original molecular structure of the rock.

metatarsals: foot bones.

metoposaurs: large aquatic amphibians of Triassic age; metoposaur fossils have been found in the Newark Supergroup rocks.

mosasaurs: large extinct marine lizards of the Late Cretaceous; related to modern monitor lizards.

multituberculates: extinct small mammals with distinctively cuspate teeth, they were the most common and diverse of the mammals in the Cretaceous Period.

nautiloids: chambered-shelled cephalopods with simple chamber walls; shells may be either straight or coiled. Includes the living pearly nautilus, as well as numerous extinct forms.

nodosaurs: members of the ankylosaur family (armored dinosaurs) that did not possess an enlarged bony club at the end of the tail.

nothosaurs: small aquatic Triassic reptiles with long necks; some authorities believe they are the ancestors of the plesiosaurs.

ornithischian: belonging to the order of dinosaurs distinguished by a rear-pointing extension of the pubis bone in the hip and a predentary bone in the lower jaw; bird-hipped dinosaurs. Includes armored and plated dinosaurs (thyreophorans), marginocephalians (horned dinosaurs and their relatives), and ornithopods.

ornithopods: a large group of ornithischian dinosaurs including the hypsilophodonts, the iguanodontids, and the hadrosaurs. The group displays increasing specialization of the jaws for grinding and chewing tough vegetation.

ornithosuchian: a group of small, lightly built Triassic archosaurs of thecodont grade, among which probably lies the ancestry of the dinosaurs.

orogeny: an episode of mountain building.

ostracoderms: primitive jawless fish of the earlier part of the Paleozoic Era.

paleontology: the study of the remains or traces of prehistoric life.

paleopathology: the study of the evidence for injury and disease in prehistoric organisms.

Pangaea: the supercontinent incorporating most of the earth's landmasses, lasting from Late Paleozoic to middle Mesozoic time.

Permo-Triassic mass extinction: the disappearrance of numerous Paleozoic organisms from the fossil record, mostly marine forms, but also involving mammallike reptiles on land. This event marks the end of the Paleozoic Era and the beginning of the Mesozoic Era. It is generally dated at around 245 million years ago.

phytoplankton: floating marine plants, mostly microscopic; the base of the food pyramid in the ocean.

phytosaurs: extinct aquatic archosaurs similar in appearance to crocodiles but distinguishable from them by the position of the nasal openings just in front of and between the eyes.

plate tectonics: study of the interactions between the large, slowly moving plates that make up the crust of the earth. These interactions are responsible for various geologic phenomena such as earthquakes, volcanoes, and mountain-building.

playa: expanding and contracting lake, usually found in arid regions.

plesiosaurs: medium to large marine reptiles of Jurassic and Cretaceous age; heavy-bodied with four flippers and a short tail, some forms had a very long neck with a small head (elasmosaurs), while other types had a long head and a short neck (pliosaurs).

pleurodires: side-necked turtles that withdraw their heads into their shells by flexing their necks sideways. Once more widely distributed over the globe, they are now restricted to the southern continents.

pliosaurs: short-necked plesiosaurs.

predentary: toothless bone in the front of the lower jaw shared by all ornithischian dinosaurs.

Principle of Faunal Succession: the principle of stratigraphy which states that each sedimentary layer has its own unique assemblage of contained fossils, and that the changes in the sequence of these fossil assemblages is regular and predictable.

Principle of Superposition: the principle of stratigraphy which states that in an undisturbed stack of sedimentary rocks, the rocks on the bottom were deposited before the sediments on top; hence the rocks toward the bottom are older than the rocks at the top.

protosuchids: ancestral stock of the crocodilians.

pterosaurs: the winged flying reptiles of the Mesozoic Era; related to dinosaurs but not true dinosaurs themselves.

pycnodonts: Mesozoic bony fish characterized by blunt teeth and deep bodies; their fossil teeth are sometimes called "bean teeth," and they were probably shell-crushers.

radioisotope: a radioactive form of an element; isotopes of the same element are all the same chemically, but have different atomic weights because of differing numbers of neutrons in the nuclei of their atoms. Radioisotopes are useful in determining geologic ages of rocks because of the unique half-life decay rate at which the radioactive isotope transforms into a stable isotope.

raptors: technically, birds of prey such as hawks; now commonly used to refer to the smaller large-clawed theropods such as *Deinonychus.*

rauisuchids: quadrupedal archosaurs of thecodont grade, they were the largest land predators of the Triassic Period.

rudists: large reef-forming clams of the Cretaceous tropical seas.

saurischian: order of dinosaurs distinguished by having a pubis (pelvic) bone that points forward and down, and by the possession of an asymmetrical hand. Includes theropods (meat-eating dinosaurs) and sauropods.

sauropods: the large, four-legged, heavy-bodied, plant-eating, saurischian dinosaurs with small heads, long necks, and long tails. Includes such famous giant dinosaurs as *Apatosaurus, Diplodocus, Brachiosaurus,* and *Seismosaurus.*

sedimentary: referring to the rocks formed by the processes of weathering, erosion, and deposition. Usually these are sediments deposited in layers, or strata.

semionotid: primitive bony fish sometimes found in "fish kill" layers of the Newark Supergroup playa lake deposits.

sphenodontians: lizardlike reptiles, largely extinct except for the modern tuatara.

stratigraphy: the study of layered rocks, primarily sedimentary rocks.

stromatolite: cabbage-shaped mounded build-ups of carbonate material made by blue-green algae in shallow tidal environments.

stromatoporoids: mounded structures thought to have been made by sponges and/or algae; major reef formers in the Middle Paleozoic seas.

synapsids: members of the reptilian subclass Synapsida, characterized by a skull with a single temporal opening behind the eyes. Includes all the mammal-like reptiles of Permian and Triassic age.

taphonomy: literally, "burial laws"; the subdiscipline of paleontology that studies how living organisms become fossils.

teleosaurs: marine crocodiles of the Mesozoic Era.

Tethys Sea: Mesozoic seaway that extended equatorially between the northern continents and the southern landmasses; named after the titaness Tethys, wife of Oceanus in Greek mythology.

tetrapod: vertebrates with four limbs, including all land animals and their descendants who returned to the sea. Does not include fishes.

thecodont: the condition of the jaw when it contains holes or sockets out of which the teeth grow; also, the evolutionary grade of possessing socketed teeth.

theropods: meat-eating saurischian dinosaurs, including *Dryptosaurus.*

tillodonts: a small group of extinct Paleocene and Eocene mammals, about the size of bears, possessing claws but thought to have fed on tough plant material because of their long incisors.

Triassic: first period of the Mesozoic Era, lasting from 245 million to 208 million years ago. Named from the three-part division of the rocks of this age seen in Germany.

trilobite: extinct group of marine arthropods whose name means "three-lobed," referring to the longitudinal division of the animals' body into three well-defined lobes.

trionychids: soft-shelled turtles.

tuatara: only surviving representative of the ancient sphenodontians; found on a few islands in New Zealand, it is a true living fossil.

zooplankton: marine floating animals, many of which are quite small; an important link in the ocean food chain.

Annotated Bibliography

There are many dinosaur books on the market, and the following list is not exhaustive. You may find some of these titles helpful as reference works for further study.

ALEXANDER, R. M. 1989. *Dynamics of Dinosaurs and Other Extinct Giants.* New York: Columbia University Press. Interesting biomechanical approach to such problems as dinosaur speed, posture, metabolism, etc.

BAKKER, R. T. 1986. *The Dinosaur Heresies.* New York: William Morrow & Co. Bakker's provocative ideas on dinosaur warm-bloodedness and behavior.

BIRD, R. T. 1985. *Bones for Barnum Brown.* Fort Worth, Tex.: Texas Christian University Press. Captures the flavor of old-time dinosaur hunting in the American West.

DESMOND, A. J. 1975. *The Hot-Blooded Dinosaurs.* New York: Dial Press. A nice historical treatment that also summarizes arguments for dinosaur endothermy.

DINGUS, L., AND OTHERS. 1995. *The Halls of Dinosaurs: A Guide to Saurischians and Ornithischians.* New York: American Museum of Natural History. Museum guidebook for the new dinosaur exhibits at the American Museum.

DIXON, D., AND OTHERS. 1988. *The Macmillan Illustrated Encyclopedia of Dinosaurs and Prehistoric Animals.* New York: Macmillan Publishing Co. Profusely illustrated. Covers all vertebrate classes.

CARPENTER, K., AND CURRIE, P. J., EDS. 1990. *Dinosaur Systematics.* New York: Cambridge University Press. A collection of technical papers on dinosaur classification.

CARPENTER, K., HIRSCH, K. F., AND HORNER, J. R., EDS. 1994. *Dinosaur Eggs and Babies.* New York: Cambridge University Press. A collection of scientific papers on dinosaur nests and juveniles.

COLBERT, E. H. 1968. *The Great Dinosaur Hunters and Their Discoveries.* New York: Dover Publications. A classic history of dinosaur paleontology.

GAFFNEY, E. S. 1990. *Dinosaurs.* New York: Golden Press. One of the Golden Guide series; small and inexpensive but a good introduction to dinosaur classification.

GILLETTE, D. G. 1994. *Seismosaurus: The Earth Shaker.* New York: Columbia University Press. Nontechnical account of the discovery, excavation, and study of the longest dinosaur now known.

FARLOW, J., ED. 1989. *Paleobiology of the Dinosaurs.* Boulder, Colo.: Geological Society of America. Technical papers on dinosaur biology.

HORNER, J. R., AND GORMAN, J. 1988. *Digging Dinosaurs.* New York: Workman Publishing. Focuses on Horner's discovery of dinosaur nests, eggs, and babies in western Montana.

HORNER, J. R., AND LESSEM, D. 1993. *The Complete T. rex.* New York: Simon and Schuster. Everything you always wanted to know about the last and largest of the great meat-eating dinosaurs.

JACOBS, L. 1995. *Lone Star Dinosaurs*. College Station, Tex.: Texas A&M University Press. Nicely illustrated account of dinosaur digs in Texas.

KIELAN-JAWOROWSKA, Z. 1969. *Hunting for Dinosaurs*. Cambridge, Mass.: MIT Press. Interesting account of the Polish-Mongolian expeditions to the Gobi Desert.

LESSEM, D. 1992. *Kings of Creation*. New York: Simon and Schuster. Discusses some of the latest discoveries in dinosaur paleontology, and the scientists who are making them.

LOCKLEY, M. 1991. *Tracking Dinosaurs*. New York: Cambridge University Press. Dinosaur footprint expert discusses dino tracks.

LUCAS, S. G. 1994. *Dinosaurs: The Textbook*. Dubuque, Mont.: William. C. Brown. Introductory college textbook for a course on dinosaurs.

MCGOWAN, C. 1991. *Dinosaurs, Spitfires, and Sea Dragons*. Cambridge, Mass.: Harvard University Press. Good anatomical treatment of extinct reptiles.

NORMAN, D. 1985. *The Illustrated Encyclopedia of Dinosaurs*. New York: Crescent Books. Already a bit dated, but worth it for the graphics.

NORMAN, D. 1991. *Dinosaur!* New York: Prentice Hall. Companion volume to the TV series of the same name; nicely illustrated.

PREISS, B., AND SILVERBERG, R., EDS. 1992. *The Ultimate Dinosaur*. New York: Bantam Books. A collection of articles by dinosaur experts, coupled with stories about dinosaurs by science fiction writers.

RUSSELL, D. 1989. *An Odyssey in Time: The Dinosaurs of North America*. Minocqua, Wis.: Northword Press. Lavishly illustrated and well written by a respected Canadian paleontologist.

WEISHAMPEL, D. B., DODSON, P., AND OSMOLKA, H. 1990. *The Dinosauria*. Berkeley: University of California Press. The single most comprehensive technical treatment of the dinosaurs.

WILFORD, J. N. 1988. *The Riddle of the Dinosaur*. New York: Alfred A. Knopf. Easy to read historical treatment that covers modern debates over warm-bloodedness and dinosaur extinction.

INDEX

(numbers in italics refer to illustrations)

Abbott, C. C., 67
Academy of Natural Sciences: and early dinosaur discoveries, 28–31, 34, 35, *36*, 37; dinosaur specimens from New Jersey, *96, 98, 104, 105, 108, 122;* fossil mammal specimens from New Jersey, 131, 143, 147
Acadian orogeny, 18, 20
acetabulum, 22–23
Aftonian interglacial, 137
Agerostrea, 75, *A. falcata, 79*
Albertosaurus, 67
Alexandria, 46
alligators, 22, 61, 113
Allosaurus, 59
Alphadon, 114
Alvarez, Walter, 117
Alvarez, Luis, 117
amber, 63–67
American Museum of Natural History, 38, 66, 68, *100, 107,* 141, 147
American Philosophical Society, 28, 59, 92, 94
Ammodon leidyanus, 136
ammonites, 78, 80, *87,* 115, 122, 123, 124, 125
amphibians, 19–20, 113
Anaklinoceras, 80
anapsids, 45
Anchippodus riparius, 131
Anchisauripus, 49, 50, 52–53
Anchitherium, 137
Anchura pennata, 77
angiosperms, 62–63

ankylosaurs, 25, 97
Anomaeodus phaseolus, 81
anthracite, 20
Antrodemus, 59
Apatopus, 49
Apatosaurus, 59
Appalachian Mountains, 4, 14, 60
Appalachian Trail, 15–16
Archelon, 83
archosaurs, 21–26
armored dinosaurs, 25, 61, 97–98, 110
Arney's Mount, 2
Arthrophycus, 16
Arundel Formation, 61
asteroid impact, 55–56, 119–121
Astrodon, 61
Atractosteus, 81
Atreipus, 49
Atlantic Highlands, 2, 88
Avondale, 151

Baird, Donald, 28, 54, 67, 68, 83, 91
Bakker, Robert, 62–63, 106
baleen whales, 134
Baltimore Canyon trough, 60
Barnsboro, 35, 36, 37, *104, 106, 108, 109*
basalt, 43, 51
Basilosaurus, 131
Batrachopus, 49
bears, 140
beavers, 140, 142
Beekmantown Formation, 14
Belemnitella americana, 79
belemnites, 80

Belvidere, 139
bentonites, 14
Bergen, 45
Bergen Museum of Art and Science, 147
Berry, Joyce, 40
Big Badlands, 132
Big Brook, 82, 97
birds, 32; Cretaceous, 61, 65, 87, 113; similarity to dinosaurs, 32, 105–106, 127–128; Tertiary, 130–131
bird-mimic dinosaurs, 99, *101*
bison, 143
bivalves, 74–75, 124
Brontotherium, 132
Blackwood, 136, 141
Blairstown, 14, 142
bobcats, 140
bone wars, 37–38, 59
Boonton, 55, 56
Boonton Formation, 43, 55, 56
Bossardville Limestone, 16
Bothremys, 83
brachiopods, 4, *12,* 14, 17, 77, 124–125, *133*
brachiosaurs, 59–60, 61
Brachiosaurus, 59
Brachycheirotherium, 46
Brown, Barnum, 67–68, 141
bryozoa, 17, 125
Buckland, William, 7, 27
bulldog tarpon, *81,* 82
Burlington County, 63, 67, 85, 98, 99, 102

caliche, 48
Calvert Cliffs, 132, 134, 137
Calvert Formation, 134
Camarasaurus, 59
Cambrian, 9, 10–14, 15
Camden County, 29, 30, 94, 136, 141
camel family, 136
Cape May, 1, 143
Cape May Formation, 132
carbon-14, 7
Carboniferous, 19
Carcharodon megalodon, 134, *135*
caribou, 143
Carpentersville, 14
Casteroides ohioensis, 142
Catskill beds, 19

Cenozoic, 5, 129–146
cephalopods, 14, 76, 78, *79*
ceratopsians, 25, 114
Cervalces scotti, 142, 143
cetaceans, 134
chalk, 74, 117, 118
chambered nautilus, 76, 78
cheetah, 140
Cheirotherium, 46, 49
Chestnut Run, 109
Chicxulub, 121
Choristothyris plicata, 77
Cimoliasaurus, 85
clams, 17, 74–76, *79, 133*
Clidastes, 85
Cliffwood Beach, 64
Cliona cretacica, 77
coal, 20
coccolithophorids (coccoliths), 74, 114–115, 117
coelacanths, 43
Coelosaurus antiquus, 98, 100
Cohansey Formation, 1, 132
Colbert, Edwin, 106
collecting fossils, 149–154
Columbian mammoths, 142
conifers, 62
continental drift, 18, 138
continental shelf, 60, 142–145
Cook, George, 98
Cope, Edward Drinker: and early dinosaur discoveries, 34–39, 59, 84, 92–93, 95, 99, 103, 105–106, 108, 109; Cope's Law, 127; and entelodonts, 136
Cope's Law, 127
corals, 11, *12,* 14, 17, 124, 125, 126
Corson's Inlet, 154
coyotes, 140
crabs, 80
craters, 56, 119, 120–121
Cream Ridge, 2
Cretaceous, 5, 55, 56, 60–69; marine deposits, 2, 71–88, 89–90; Late Cretaceous dinosaurs, 67–69, 90–111
Cretaceous-Tertiary (K/T) boundary, 113–129, 131
Cretolamna appendiculata, 81
crinoids, *12,* 17, *133*
crocodiles, 21, 22, 24, 26, 29, 35; early

forms, 44, 49, 55; Cretaceous, 83–84, 95; as K/T survivors, 113, 114, 126; Paleocene, 129
crossopterygians, 19
crustaceans, 80
Crystal Palace Exhibition, 7, 31
cuesta, 2
Cumberland County, 2, 134
Cuvier, Baron Georges, 5–6, 116
cycads, 62
cynodonts, 21

Decker Fm., 16
deer, 140, 142
Delaware Water Gap, 15, 17, 153
Deinodon, 35, 103
Deinonychus, 38–39
Deinosuchus, 83–84
Denton, Robert, 103
deoxyribonucleic acid (DNA), 66–67
Devonian, 17–19
diapsids, 21, 26
Diatryma, 131
Diceratherium, 136
Dillon, Penney, 64
Dilophosaurus, 51
dinosaurs: early discoveries in England, 7; ancestors of, 20, 21, 22; hip structure, 22–23; dinosaurs defined, 24–25; diversity of, 25–26; early discoveries in New Jersey, 27–40; in Newark Basin, 43, 46–57; Jurassic, 59–60; Early Cretaceous, 60–61; and flowering plants, 62–63; cloning dinosaurs, 66–67; in Raritan Formation, 67–69; as prey for giant crocodiles, 83–84; Late Cretaceous, 67–69, 89–111; metabolism, 106; baby, 107; paleopathology, 109–*110*; extinction, 111, 113–114, 122–123, 127–128, 129; footprints, 149–157; fossils in stream gravels, 95–97, 154
Diplodocus, 59, 60
Diplotomodon horrificus, 104, 110
dog family, 137
Draco, 45
Dryptosaurus, 110, 111, 122; *Dryptosaurus aquilunguis*, 38–39, 97, 99, 102–108

elasmosaurs, 84, 85

Elasmosaurus, 37
elephants, 5, 140
elk-moose, 142
Enchodus, 81, 82, 92, 115, 122
endangered species, 127, 146
Englishtown Formation, 75
entelodonts, 136
Eocene, 131, 132, 133
Eoraptor, 24, 41
Eryops, 20
Etea delawarensis, 133
Eubrontes, 51, 52, *54*, 149
eurypterids, 15
Exogyra, 75, 124; *E. cancellata*, 79
exogyrines, 115

Felder, Larry, 55
Feltville Formation, 43, 51
fish, 11, 16, 19; in Newark Basin, 43, 55; Cretaceous, 82, 87, 88, 95; and K/T boundary, 115, 122, 124, 126
Florham Park, 151
flowering plants, 62–63, 66–113
footprints: in Newark Basin, 43, 46–56, 149–151; Cretaceous, 67–69; collecting, 149–150; photography, 150; mapping, 150; rubbings, 150–151; sites, 151
foraminifera (forams), 74, 115, 117, 125
Fort Lee, 44
Foulke, William Parker, 29–32
foxes, 140
Franklin, Benjamin, 28
Frenchtown, 151

gastroliths, 61
gastropods, 76
geomagnetic reversals, 116–117, 118
ghost shrimp, 80
giant ground sloth, 140, 142, 143, 154
Gibbsboro, 136
Gilmore, Edward, 65
glauconite, 71, 123
Globidens, 85
Gloucester County, 28, 35, *36*, *37*, 85, 86, 87, 94, 103, 104, 110, 122, 123, 126, 136
Grallator, 48, 49, 52
Grandstaff, Barbara, 102, 143
Granocardium tenuistriatum, 79

Granton Quarry, 45
graptolites, 14–15
Graywacz, Katherine, 61, 67
Great Swamp Outdoor Education Center, 147
greensand, 71, 76, 82, 83, 89, 98, 123, 124
Gryphaeostrea vomer, 133
gymnosperms, 62
Gyrodes abyssinus, 77

Haddonfield, 29, 30, 35, 36, 40
Haddon Township, 40
hadrosaurs, 25, 29–35, 92–97, 106–109, 111, 114, *122*
Hadrosaurus foulkii, 25; discovery of, 29; description of, *30*–31; skeletal reconstructions, *32–33,* 34, 35, 39–40, 59, 90–92, 93, 95, 111
Hadrosaurus minor, 106–107, 111, *122*
Halisaurus, 87
Hardyston Formation, 13
Hatcher, John Bell, 60
Haverford College, 34–35
Hawkins, Benjamin Waterhouse, 7, 31
Hayden, Ferdinand, 28, 59, 91
Hell Creek Formation, 120, 123, 127
Hemipristis serra, 135
Herrarasaurus, 41
Hesperornis, 113
heteromorph ammonites, 80
Hexanchus, 135
High Falls Formation, 16
Highlands, 4, 9–13
High Point, 1, 15
Hildebrand, Alan, 120–121
Hillsborough, 46
Holgate, 154
Hook Mountain Basalt, 55
Hopkins, John, 29
Horner, John, 83, 91
Hornerstown Fm., *122, 123,* 124, 125, 126, 129, 132
horses, 137, 143
Hunterdon Hills, 51
Hutton, James, 116
Huxley, Thomas Henry, 32
hybodonts, 80, *81,* 115
Hybodus, 81
Hylaeosaurus, 27

Hylonomus lyelli, 20
Hypsognathus fenneri, 45–46, 47, 55

Icarosaurus siefkeri, 45
Ice Age, 1, 137–146, 154
Ichthyornis, 113
Iguanodon, 7, 27, 28, 30, 61, 91, 106
Illinoisan glaciation, 137
Imperial mammoth, 142
inoceramids, 74, 115
insects, 19, 20, 22, 63–66, 113, 139
interglacial phases, 137, 145
iridium, 117–119, 121, 124
Ischigualasto Formation, 41
Island Beach State Park, 154
Isurus desorii, 135
Isurus hastalis, 135

Jacksonburg Fm., 14
jaguar, 140
jaguarundi, 140
javelina, 136
jawless fish, 15, 16
Jefferson, Thomas, 145
Jericho, 134
Johnson, Christopher, 61
Johnson, Meredith, 67
Johnson, Ralph, 64, 97
Judith River beds, 91, 127
Jurassic, 42, 51–56, 59–60, 82, 83, 84, 149

Kansan glaciation, 137
karst topography, 139
Kinkora, 63
Kirkwood Formation, 132, 134, 136
Kittatinny Ridge, 13, 15–16
Kittatinny Valley, 13
Knight, Charles R., 38–39
Komodo Dragon, 85
K/T boundary, 113–128, 129, 131

LaBrea tarpits, 140–141
Laelaps aquilunguis, 35–36, 38–*39,* 99, 103–106, 108, 110
Laelaps macropus, 99, *100,* 108, 110
Lagosuchus, 22
lamnoid sharks, 80
Larson, John, 143, 144
Laskowich, C., 54

Lauginiger, Edward, 102
lava flows, 50, 51, 55
lizards, 21, 22, 24, 26, 85, 113, 114
Lea, Isaac, 29
Leidy, Joseph: and *Hadrosaurus foulkii*, 28–32; and E. D. Cope, 34; and *Antrodemus*, 59; and Cretaceous dinosaurs from New Jersey, 90, 91, 92, 94, 110; and Tertiary mammals, 131, 132, 136
Lewis and Clark, 145
Liberty Township, 141
Lincoln Park, 53
Liodon, 87
Little Falls, 151
lobsters, 80
Lockatong Formation, 43–45, 46, 48
Lockwood, Samuel, 92
Long Beach Island, 154
Lyell, Charles, 116, 131

Magothy Formation, 63, 64, 66, 67, 71, 75
mammals, 21, 55; and K/T boundary, 114, 115, 128; Tertiary, 129, 131–132, 134, 136–137; Pleistocene, 140–146
mammal-like reptiles, 21, 55
mammoths, 141–142, 143, *144*
Mammut americanum, 140
Manacouagan crater, 56
Manalapan, 107
Manasquan Formation, 132, 133
Manasquan River, 134, 136
manatees, 143–144
Manson Crater, 120
Mantell, Gideon, 7, 27
Mantua Township, 86, 109, 110, 122, 126
Maple Shade, 75, 85
Marasmius, 66
marginocephalians, 25, 26
marl, 2; and early dinosaur discoveries, 29, 35–39; as marine deposits, 71, 76, 78, 89; Cretaceous dinosaur discoveries in, 95, 106, 109; fossil birds in, 113; fossil mammals in 131, 134, 136
Marlboro, 97, 99, 100
Marsh, Othniel Charles: rivalry with Cope, 34–38, 59, 60; and Cretaceous dinosaur discoveries in New Jersey,

92, 103, 106, 107, 109; and fossil mammals from New Jersey, 136
Martin, Paul, 145
Martinsburg Formation, 14
Marshalltown Formation, 75, 97, 99
marsupials, 115
mass extinctions, 6, 17; Permo-Triassic, 21, 113, 125; Triassic-Jurassic, 50–51, 55–56; Cretaceous-Tertiary (K/T), 113–128; timing of, 115, 119; terrestrial causes, 115–116; extraterrestrial causes, 116–121; Alvarez hypothesis, 119–121, impact tsunami, 121; Pleistocene, 145–146
mastodons, 140–*141*, 142, *143*, 145, 154
Matawan Group, 75, 85, 94
Meadowlands Museum, 147
Megalonyx wheatleyi, 140
Megalosaurus, 7, 27, 28, 35, 106
Merchantville, 94
Merchantville Formation, 74–75, 78, 95
Mesozoic, 21–25, 55–113, 129
Metasequoia, 62
metoposaurs, 44, 55
Middlesex County, *64*, 67, 71, 85
Middletown, 97, 153
Milankovitch cycles, 138
Milford, 41, 151
Millington, 151
Miocene, 2, 131, 132, 134–137
Mississippian, 20
Monmouth County: 2; Cretaceous marine fossils, 76, 82, 85; Cretaceous dinosaur fossils, 88, 92, 95, 97, 99, 100, 102, 107, 108; fossil mammals, 131, 134, 137, 142
Monmouth Group, 75, 85, 121
Morris Museum of Arts and Sciences, 147
Morrison Formation, 38, 59–60, 62–63
mosasaurs, 6, 35, 85–87, 88, 90, 115, 122, 124, 126, 131
Mosasaurus, 6; *Mosasaurus maximus*, 86, 87
Mount Holly, 2
Mount Hope, 142
Mount Laurel, 102
Mount Laurel Formation, 75, 76, 80
Mullica Hill, 2, 103, 110
multituberculates, 115

mushrooms, 66
musk ox, 143
mussels, 74

nautiloids, 14, 17
Navesink Fm., 75, *76*, 80, 86, 87, 88, 104, 109, 110, 122, *123*
Nebraskan glaciation, 137
Neshanic Station, 46
Newark, 151
Newark Basin, 4, 41–56, 149, 151
Newark Museum, 148
Newark Supergroup, 4, 42–56, 151
New Brunswick, 2
New Jersey Geological Survey, 67, 94
Newton, 14
New Jersey State Museum, *32*, 44, 45, 47, 52–54, 67–68, *86*, 97, *101*, 102, 126, 141, 143, 144, 145, 147
New Vernon, 151
Niobrara Formation, 74
nodosaurs, 61, 97–98, 110
nothosaurs, 84
neutron activation analysis, 118

octopus, 14, 76
Odontaspis elegans, 135
odontocetes, 134
Ogden, David, 94
Oleneothyris harlani, 124, 133
Oligocene, 131–132
Olsen, Paul, 52, 55
Ordovician, 14–15
ornithischians, 23–24, 26, 49
Ornithomimus antiquus, 98, 99
ornithopods, 25–26, 61
ornithosuchians, 22
Ornithotarsus immanis, 93, 95, 110
ostracoderms, 16
Ostrom, John, 38
Otodus obliquus, 129
otters, 140
Owen, Richard, 7, 27, 28, 30, 31
oysters, 74, 75–*76, 79*, 82, 124, 125

pachycephalosaurs, 25
Paleocarcharodon, 129
Paleocene, 5, 123, 125–126, 129–131, 132, 133
Paleophis, 129–130

Paleozoic, 4, 9–12, 21, 125, 139, 153, 154
Paleozoic Museum, 31
Palisades, 51
Pangaea, 24, 41, 51, 56–57, 60, 116
Parris, David, 145
Passaic, 46, 47
Passaic Formation, 43, 48–51, 54, 55
Peabody Museum of Natural History, 34, 36, 38, 61, 68, 93, 97, 107, 149–150
Peale, Charles Wilson, 28
Peale Museum, 28
Peapack, 14
peat deposits, 140
peccary, 136, 140
pelican, 88
Penfield, Glen, 120–121
penguins, 84
Pennsauken Formation, 137
Pennsylvanian, 20
Pentacrinus bryani, 133
Permian, 5, 21
Permo-Triassic mass extinction, 21, 113, 125
petrified wood, 136
pleurodires, 83
Potomac Group, 60, 62, 67
phytoplankton, 74
phytosaurs, 22, 44, 45, 49, 55
pika, 140
Pinelands, 1–2, 154
Pitman, 109
Placenticeras, 78
plankton, 74, 115, 117, 118, 125
Pleistocene, 5, 132, 137–146
plesiosaurs, 37, 38, 84–85, 90, 103, 115, 124, 126
Pliocene, 5, 137
Plioplatecarpus, 87
pliosaurs, 84, 85
plutonium, 117, 118
Point Pleasant, 154
pollen, 50–51, 62, 63
Pompton Falls, 151
Poricy Brook, *76*, 82, 97, 153–154
porpoises, 134
Port Kennedy cave deposit, 139–140
Postosuchus, 22
Poxono Island Formation, 16
Precambrian, 4, 9
Priconodon, 61